Springer Theses

Recognizing Outstanding Ph.D. Research

Aims and Scope

The series "Springer Theses" brings together a selection of the very best Ph.D. theses from around the world and across the physical sciences. Nominated and endorsed by two recognized specialists, each published volume has been selected for its scientific excellence and the high impact of its contents for the pertinent field of research. For greater accessibility to non-specialists, the published versions include an extended introduction, as well as a foreword by the student's supervisor explaining the special relevance of the work for the field. As a whole, the series will provide a valuable resource both for newcomers to the research fields described, and for other scientists seeking detailed background information on special questions. Finally, it provides an accredited documentation of the valuable contributions made by today's younger generation of scientists.

Theses are accepted into the series by invited nomination only and must fulfill all of the following criteria

- They must be written in good English.
- The topic should fall within the confines of Chemistry, Physics, Earth Sciences, Engineering and related interdisciplinary fields such as Materials, Nanoscience, Chemical Engineering, Complex Systems and Biophysics.
- The work reported in the thesis must represent a significant scientific advance.
- If the thesis includes previously published material, permission to reproduce this must be gained from the respective copyright holder.
- They must have been examined and passed during the 12 months prior to nomination.
- Each thesis should include a foreword by the supervisor outlining the significance of its content.
- The theses should have a clearly defined structure including an introduction accessible to scientists not expert in that particular field.

More information about this series at http://www.springer.com/series/8790

Peter Shadbolt

Complexity and Control in Quantum Photonics

Doctoral Thesis accepted by
the University of Bristol, UK

Springer

Author
Peter Shadbolt
Imperial College London
London
UK

Supervisor
Prof. Mark Thompson
H. H. Wills Physics Laboratory, Center for Quantum Photonics
University of Bristol
Bristol
UK

ISSN 2190-5053 ISSN 2190-5061 (electronic)
Springer Theses
ISBN 978-3-319-21517-4 ISBN 978-3-319-21518-1 (eBook)
DOI 10.1007/978-3-319-21518-1

Library of Congress Control Number: 2015947102

Springer Cham Heidelberg New York Dordrecht London

Printed on acid-free paper

Springer International Publishing AG Switzerland is part of Springer Science+Business Media
(www.springer.com)

To my parents

Supervisor's Foreword

Since 2001, it has been known that a universal quantum computer can be built using single photon sources, single photon detectors, and an adaptive interferometer. There has since been a great deal of interest in practical implementations of such a machine. Early experiments were performed using bulk-optical elements: macroscopic beamsplitters, lenses and filters bolted to an optical bench. However, a useful, universal device will require many thousands of components, and can only be realised in an integrated, lithographically fabricated platform. Over the past decade, the capability of integrated photonics has grown to meet the needs of the classical telecommunications and computing industries, to the extent that lasers, filters, switches, and optical interconnects can all now be implemented in planar waveguide chips using standard semiconductor fabrication techniques.

Proof-of principle experiments, performed over the last few years, have demonstrated that this hardware is suitable for applications in quantum photonics. Although the overarching goal is the development of practical quantum technologies, these experiments also provide tangential benefits, enabling tests of foundational quantum mechanics with a degree of complexity and control which would be extremely challenging—if not impossible—using bulk-optical apparatus.

Pete Shadbolt's thesis captures both sides of this coin. His work showcases the utility of integrated photonics for optical quantum computing, demonstrating novel on-chip implementations of quantum state preparation, logic gates, quantum algorithms, and state characterization methods. Meanwhile, he has used the same reconfigurable hardware to explore foundational quantum physics, including complementarity and nonlocal quantum correlations.

A large fraction of Pete's thesis makes use of a reconfigurable two-qubit silica-on-silicon chip. Pete has shown how this device can be used to generate and characterise Bell states, perform quantum process tomography, and violate Bell tests. While these experiments have previously been performed in bulk optics, these were the first on-chip demonstrations of these essential protocols. Pete went on to use this chip to test a new method which guarantees Bell violation in the absence of a shared reference frame, a situation which can occur due to polarization rotations in optical fiber, for instance. He describes an extension of this scheme which greatly facilitates the characterisation of entangled photon sources in the presence of realistic experimental noise.

The same device was used to implement a novel twist on Wheeler's classic delayed choice experiment, testing the counterintuitive wave-particle duality of the photon. In this new experiment, Wheeler's binary "choice" was replaced by a qubit, allowing wave and particle behaviours to be simultaneously observed in coherent superposition.

With respect to applications, Chap. 5 of Pete's thesis describes the experimental implementation of a new and unorthodox algorithm for quantum chemistry on a quantum computer. This algorithm has potential benefits in terms of the depth and complexity of the quantum circuit required to solve the electronic structure problem using a quantum computer.

In Chap. 6 of this thesis, the experimental complexity is significantly increased. Pete describes the development of a 16-channel multiphoton detection system, used in a series of experiments which examine quantum interference in waveguide arrays. These devices include continuously coupled quantum walks, and a Haar-random unitary circuit. Using up to 5 photons in 21 waveguides these experiments access a Hilbert space of around 50,000 dimensions, providing insight into the challenges associated with state characterization and algorithmic verification in large-scale quantum devices.

This thesis is characterised by the breadth of topics covered, from practical time-correlated single photon counting through to quantum algorithms and foundational quantum theory. It describes a large number of new results from an experimental standpoint, with extensive, practically relevant detail. As such, I believe that it will be an invaluable addition to the library of any student of experimental integrated quantum photonics.

Bristol, UK
September 2015

Prof. Mark Thompson

Abstract

Quantum mechanics predicts phenomena which have no classical analogue. This modifies our understanding of the capability of physical machines. Single photons, together with simple interferometers and single photon detection have been shown to be universal for the construction of many such machines. The nascent field of integrated quantum photonics addresses the scalability and practicality of such machines, and their integration in miniaturized monolithic chips.

In this work, we explore the scope and flexibility afforded by integrated quantum photonics, both in terms of practical problem-solving, and for the pursuit of fundamental science. We demonstrate and fully characterize a two-qubit quantum photonic chip, capable of arbitrary two-qubit state preparation. We make use of the unprecedented degree of reconfigurablility afforded by this device to implement a novel variation on Wheeler's delayed choice experiment, and test a new technique to obtain nonlocal statistics without a shared reference frame. We demonstrate a new algorithm for quantum chemistry, simulating the helium hydride ion. Finally, we demonstrate multiphoton quantum interference in a large Hilbert space, and discuss implications for computational complexity.

Acknowledgments

I have been enormously privileged to spend the past 4 years of my life studying physics in Bristol. I owe this privilege Jeremy O'Brien and Mark Thompson, who first gave me this opportunity, and who have given me supervision, advice, and freedom throughout.

Beyond that there are simply too many people to thank. Hundreds of people have helped me out. Please believe me when I say that I am grateful for all of the selfless advice, teaching, time in the lab, proof reading, comradeship, and good conversation which I have always enjoyed.

Contents

Acronyms

APD	Avalanche photodiode
API	Application protocol interface
BB84	Bennett & Brassard 1984
BBO	β-barium borate (β-BaB_2O_4)
BD	Beam dump
BiBO	Bismuth borate (BiB_3O_6)
BS	Beamsplitter
CHSH	Clauser-Horne-Shimony-Holt
CNOT	Controlled-not
CNOT-MZ	Reconfigurable two-qubit chip
CNOT-P	Postselected linear-optical CNOT
CPU	Central processing unit
CW	Continuous-wave
CZ	Controlled-Z
DAC	Digital-to-analog converter
DC	Directional coupler
DI-QKD	Device-independent quantum key distribution
DM	Dichroic mirror
ECC	Error-correcting code
ECT	Extended Church-Turing thesis
FCI	Full configuration interaction
FPGA	Field-progammable gate array
FWHM	Full-width half-maximum
GUI	Graphical user interface
HBT	Hanbury-Brown-Twiss
HOM	Hong-Ou-Mandel
HTTP	Hypertext transfer protocol
HWP	Half wave plate
IF	Interference filter
IQP	Integrated quantum photonics
KLM	Knill, Laflamme and Milburn

LOCC	Local operations and classical communication
LOQC	Linear optical quantum computing
MMI	Multimode interference
MZI	Mach-Zehnder interferometer
NV	Nitrogen vacancy
PBS	Polarising beamsplitter
PCB	Printed circuit board
PEA	Quantum phase-estimation algorithm
PMF	Polarization-maintaining fibre
QFT	Quantum Fourier transform
QKD	Quantum key distribution
QPT	Quantum process tomography
QST	Quantum state tomography
QW	Quantum walk
QWP	Quarter wave plate
RSA	Rivest-Shamir-Adleman
RTP	Room temperature and pressure
RU	Random unitary
SHG	Second-harmonic generation
SMF	Single-mode fibre
SPDC	Spontaneous parametric down conversion
SPS	Single-photon sources
TCSPC	Time-correlated single photon counting
VG	V-groove array
XOR	Exclusive-OR

Chapter 1
Introduction and Essential Physics

Krazy: "Why is Lenguage, Ignatz?"
Ignatz: "Language is that we may understand one another."
Krazy: "Can you unda-stend a Finn, or a Leplender, or a Oshkosher, huh?"
Ignatz: "No,"
Krazy: "Can a Finn, or a Leplender, or a Oshkosher unda-stend you?"
Ignatz: "No,"
Krazy: "Then I would say lenguage is that that we may mis-unda-stend each udda."

George Herriman, Krazy Kat

1.1 Introduction

Over the past century, it has become increasingly apparent that Nature, at its most fundamental level, resists analogy with human experience. Quantum theory predicts behaviour which is not explained by any classical model. As a result, we have come to understand that certain intuitive beliefs concerning the potential capability of machines do not hold. Given the ability to prepare, manipulate, and measure single quanta, there is very good evidence to suggest that we should be able to measure, communicate and compute using techniques which have no classical analogue. This new mode of operation promises enormous potential benefits in terms of speed, precision, and security.

Historically, light has played a central role both in creating and answering fundamental questions in physics. The question of the fundamental makeup of light was crucial to the development of quantum theory, and many experimental tests of the most surprising predictions of quantum mechanics were first performed using visible photons. Moreover, many of the most significant modern technologies depend entirely on the ability to manipulate and measure visible or near-visible electromagnetic radiation.

In order to implement new quantum technologies, we must choose a quantum system in which to encode information. Single photons can be readily generated and detected, and generally do not suffer from the detrimental effects of noise to the

P. Shadbolt, *Complexity and Control in Quantum Photonics*,
Springer Theses, DOI 10.1007/978-3-319-21518-1_1

same extent as other quantum particles. As such, quantum optics represents a leading approach to the implementation of almost all proposed quantum technologies.

Recently, it has been suggested that efficient and universal control over photonic quantum states could be implemented in a monolithic chip, enabling the technologies previously described. Some experimental evidence already exists to support this claim. However, in order to reach the ultimate goal of a tangible quantum advantage over classical machines, we must overcome a number of crucial challenges in photonics engineering. It is reasonable to expect that in the course of this technological development, we will, as a by-product, obtain tools which enable new science, and new understanding of quantum mechanics itself.

1.2 Thesis Outline

This chapter begins with a brief overview of quantum mechanics, entanglement, nonlocality, and prospective quantum technologies. We discuss the standard optical tool-kit in the context of quantum phenomena and quantum machines. We also discuss integrated quantum photonics. In Chap. 2, we a reconfigurable integrated photonic chip incorporating two path-encoded qubits, and show that it performs with high fidelity across a large parameter space. In the course of this work we demonstrate two-qubit quantum state and process tomography, and violate a Bell inequality on-chip. In Chap. 3, we use this device to implement a variation on Wheeler's delayed-choice experiment, showing continuous morphing between wave-like and particle-like behaviour. In Chap. 4, we consider the problem of obtaining nonlocal statistics, or certifying entanglement, without a shared reference frame. We introduce new techniques which facilitate this task, and experimentally demonstrate their feasibility. Chapter 5 introduces a new algorithm for quantum chemistry on a quantum computer, and we use this algorithm to simulate the helium hydride ion. In Chap. 6, we describe a multiphoton counting system using 16 detectors, and its application to the imaging of multiphoton quantum interference in Hilbert spaces of dimension $\sim$50,000. Chapter 7 concludes this thesis, with an outlook to future work.

1.3 Quantum Mechanics

Classical physics provides a description of the world which can be pictured in the mind's eye. The behaviour of classical objects, fields, fluids, and machines can be explained either in terms of effects which we as human beings experience and observe, or by direct and satisfactory analogy to our experience.

Over the course of the 20th century, it became increasingly evident that classical physics does not provide a complete picture of the world. In particular, two macroscopic physical effects—black body radiation and the photoelectric effect—cannot be adequately explained by a classical model. Throughout more than 100 years of discovery, Planck, Bohr, Einstein, de Broglie, Schrödinger, Dirac, and many others developed the theory of quantum mechanics, which accommodates these phenomena and predicts a great deal more. Quantum mechanics remains the most complete and accurate model of physics ever developed.

In order to construct this theory, it has been necessary to accept the existence of phenomena which resist any meaningful analogy with everyday human experience. In particular, quantum mechanics dispenses with the idea that the attributes of physical (Fig. 1.1) systems are well-defined prior to the act of measurement, as well as the notion that physics is at heart governed by deterministic processes. Quantum mechanics predicts new phenomena, such as entanglement and nonlocality, which are extreme in their departure from a common-sense understanding of the world. These effects have since been widely observed in experiments, where they are most regularly seen in nanoscale systems such as single atoms, electrons, and photons.

Very early on in the development of quantum theory, it was recognised that the surprising new effects it predicts might be used to build machines which would not be feasible in a classical model. Perhaps the most dramatic example of this was the immediate application of the new theory to the development of the atomic bomb, leading to the deaths of more than ten thousand people at Hiroshima. Quantum theory was also instrumental in the development of field-effect transistors, atomic clocks, hard disk drives, and the laser, which led to a revolution in information processing, communication, and measurement. Later on, it was suggested—by Feynman, Lloyd, Deutsch, Kitaev, and others—that coherent quantum machines, directly manipulating pure quantum states at the lowest level, might possess a fundamental advantage over their classical or semi-classical counterparts for certain tasks, including secure communication, measurement, computation, and simulation of quantum systems themselves. In contrast with the transistor, whose functionality can be reproduced by a solenoid, the capability offered by these quantum technologies would be fundamentally inaccessible to classical machines. These applications are discussed throughout this thesis.

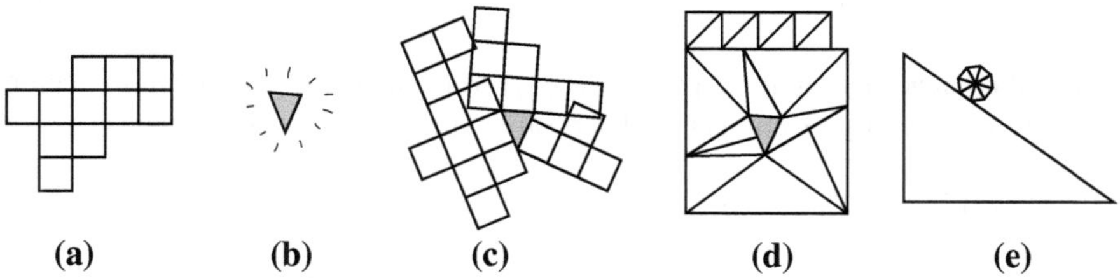

Fig. 1.1 Models of physics. **a** A *square*-based model elegantly captures the properties of many things: skyscrapers, chess boards, salt crystals. **b** However, we need only find one example—which might only be seen in a challenging or contrived experiment—to detect the incompleteness of the model. **c** □-physics does not elegantly account for the existence of triangles. **d** The new theory of △-physics is radical and unfamiliar, but it accommodates the new phenomenon well. It is arguably more elegant than the old model, and provides a more complete understanding of the world. Importantly, this new model is largely compatible with the previous understanding. **e** △-physics allows the construction of machines which are difficult to build in a □-based model

In this section I draw on notes from Nielsen and Chuang [1], Preskill [2], Dirac [3], Hannabuss [4], and Aaronson [5].

1.3.1 States

Classically, an event with n possible outcomes is described by a probability distribution $\mathcal{P}$, corresponding to a vector of n real scalars

$$\mathcal{P} = (p_1, p_2 \dots p_n); \quad p_i \in \mathbb{R}; \quad 0 \leq p_i \leq 1 \;\forall\, i. \tag{1.1}$$

Since we always obtain *some* outcome, these numbers sum to 1, i.e. the 1-norm is conserved,

$$\sum_i |p_i| = 1. \tag{1.2}$$

Quantum mechanics is the theory which naturally emerges if one attempts to replace these probabilities by complex *amplitudes*, with the condition that the 2-norm, rather than the 1-norm, is conserved

$$|\psi\rangle = (a_1, a_2, \dots a_n); \quad a_i \in \mathbb{C}; \quad ||\psi||^2 = \sum_i |a_i|^2 = 1. \tag{1.3}$$

The state of the system is completely encoded in the state vector $|\psi\rangle$, which is defined on the complex *Hilbert space*, $\mathscr{H}$. Any ray in $\mathscr{H}$ corresponds to a physical state, and two vectors represent the same state iff one is a multiple of the other. This allows construction of superposition states

$$|\psi\rangle = a_1|\psi_1\rangle + a_2|\psi_2\rangle; \quad |a_1|^2 + |a_2|^2 = 1. \tag{1.4}$$

The Hilbert space has an inner product $\langle\varphi|\psi\rangle$, which associates each pair of vectors $|\varphi\rangle$, $|\psi\rangle$ with a complex number, and is positive, linear, and skew symmetric

$$\langle\varphi|\psi\rangle \geq 0; \quad \langle\varphi|\left(a|\psi_1\rangle + b|\psi_1\rangle\right) = a\langle\varphi|\psi_1\rangle + b\langle\varphi|\psi_2\rangle; \quad \langle\varphi|\psi\rangle = \langle\psi|\varphi\rangle^*. \tag{1.5}$$

Normalization (1.3) can then be re-expressed as $\langle\psi|\psi\rangle = 1$. Volume in Hilbert space is measured by the *Haar measure* $\mathrm{d}|\psi\rangle$, which defines a notion of uniform sampling or integration over $\mathscr{H}$. In order to describe a composite system of two or more objects, Hilbert spaces are joined by means of the tensor product

$$\mathscr{H}_{AB} = \mathscr{H}_A \otimes \mathscr{H}_B; \quad |\Psi_{AB}\rangle = |\psi_A\rangle \otimes |\psi_B\rangle \tag{1.6}$$

We will often make reference to the Hilbert space *dimension d* pertaining to some physical system of interest. By this, we will usually mean the dimension of the *smallest* Hilbert space required to capture the full dynamics of the system, all things being equal. When we model a classical coin as a two-state system, we ignore many degrees of freedom—position in space, temperature etc.—which are not pertinent to the problem. Similarly, a quantum coin can be modelled as a two-state system ($|H\rangle$, $|T\rangle$) with Hilbert space dimension $d = 2$.

1.3.2 Measurements

An *observable* is a property of a physical system which can in principle be measured. Observables in quantum mechanics are described by Hermitian operators $\hat{A}$ defined on $\mathscr{H}$, which map states to states:

$$\hat{A} : |\psi\rangle \rightarrow \hat{A}|\psi\rangle; \quad \hat{A} = \hat{A}^\dagger. \tag{1.7}$$

Any observable has a spectral decomposition

$$\hat{A} = \sum_i \lambda_i \hat{\Pi}_{\lambda_i}, \tag{1.8}$$

with eigenvalues λ_i. Here, $\hat{\Pi}_{\lambda_i}$ are orthonormal *projectors* on $\mathscr{H}$, with

$$\hat{\Pi}_i \hat{\Pi}_j = \delta_{ij}; \quad \hat{\Pi}_i = \hat{\Pi}_i^\dagger. \tag{1.9}$$

If λ_i is nondegenerate, then $\hat{\Pi}_i = |\lambda_i\rangle\langle\lambda_i|$, with $\langle\lambda_i|\lambda_j\rangle = \delta_{ij}$, and $\{|\lambda_i\rangle\}$ form an orthonormal basis for $\mathscr{H}$.

When the observable $\hat{A}$ is experimentally measured, the outcome is always an eigenvalue of $\hat{A}$. The outcome of any given measurement is in general probabilistic, returning λ_i with probability

$$p(\lambda_i) = \langle\psi|\hat{\Pi}_i|\psi\rangle \tag{1.10}$$

At the time of measurement, the system is projected into an eigenstate of $\hat{A}$ corresponding to the measured eigenvalue λ_i.

$$|\psi\rangle \xrightarrow{\text{Detect } \lambda_i} \frac{\hat{\Pi}_i|\psi\rangle}{(\langle\psi|\hat{\Pi}_i|\psi\rangle)^{1/2}} = |\lambda_i\rangle\,. \tag{1.11}$$

This is the "collapse" of the wavefunction, whose interpretation remains contentious. It implies that repeated further measurements of the same operator on the same system will always yield the same eigenvalue. The expectation value of $\hat{A}$ for a state $|\psi\rangle$ is given by

$$\langle A\rangle = \sum_i p(\lambda_i)\lambda_i = \langle\psi|\hat{A}|\psi\rangle \tag{1.12}$$

The *Born rule* connects amplitudes to probabilities. It gives the probability that a system prepared in a state $|\psi\rangle$ will be detected in state $|\varphi\rangle$, as

$$p(\varphi|\psi) = |\langle\varphi|\psi\rangle|^2. \tag{1.13}$$

1.3.3 Time Evolution

Time-evolution of a classical probability distribution can be described in terms of a stochastic matrix—a matrix of real numbers whose columns each add up to one, preserving the 1-norm. Time evolution of a quantum state must preserve the 2-norm (1.3). The most general class of operators which always conserve the 2-norm of a vector on $\mathscr{H}$ are the *unitary* matrices $\hat{U}$,

$$\hat{U}\hat{U}^\dagger = \mathbf{1}; \quad \sum_i |\hat{U}_{ij}|^2 = 1. \tag{1.14}$$

Time-evolution of a closed quantum system can always be described by a unitary matrix. In the *Schrödinger picture* of quantum mechanics, we say that $\hat{U}$ evolves an input state $|\psi\rangle_{\text{in}}$ to an output state $|\psi\rangle_{\text{out}}$, as

$$|\psi\rangle_{\text{out}} = |\psi(t)\rangle = \hat{U}|\psi\rangle_{\text{in}} = \hat{U}|\psi(0)\rangle, \tag{1.15}$$

and observables $\hat{A}$ do not change as a function of time.

How is the unitary operator $\hat{U}$ connected to the physical properties of the system at hand? In general, $\hat{U}$ is generated by a Hamiltonian $\hat{H}$, according to the time-dependent Schrödinger equation

$$i\hbar\frac{\partial}{\partial t}|\psi\rangle = \hat{H}|\psi\rangle, \tag{1.16}$$

where $\hbar$ is Planck's constant. $\hat{H}$ is defined on the Hilbert space $\mathscr{H}$, and has a spectral decomposition in terms of energy eigenstates and eigenvalues, $\hat{H} = \sum_i E_i|E_i\rangle\langle E_i|$. When the Hamiltonian is fixed in time, the time-independent component of solutions of (1.16) satisfy the *time-independent Schrödinger equation* $\hat{H}|\psi\rangle = E|\psi\rangle$, where E is the energy of the state $|\psi\rangle$. The Schrödinger equation then has solutions of the form

$$|\psi\rangle_{\text{out}} = |\psi(t)\rangle = \exp\left[\frac{-i\hat{H}(t_2 - t_1)}{\hbar}\right]|\psi(0)\rangle = \hat{U}(t_2, t_1)|\psi\rangle_{\text{in}}. \quad (1.17)$$

The Hamiltonian $\hat{H}$ thus completely determines the continuous-time dynamics of the system, and can be related to the discrete-time unitary description of evolution by (1.17).

As well as the Schrödinger picture of quantum mechanics, we can equivalently adopt the *Heisenberg picture*, in which the state is thought of as remaining fixed, with observables evolving under $\hat{U}$,

$$\hat{A}_{out} = \hat{U}^{\dagger}\hat{A}_{in}\hat{U}. \quad (1.18)$$

This picture can sometimes provide a simpler analysis, especially for systems of few particles in many modes. The correspondence between these pictures can be seen as

$$|\psi\rangle_{out} = \hat{V_t}^{\dagger}\hat{U}|\psi\rangle_{in}; \quad \hat{A}_{out} = \hat{V}_t^{\dagger}\hat{A}_{in}\hat{V}_t \quad (1.19)$$

where $\hat{V}_t = \mathbf{1}$ and $\hat{V}_t = \hat{U}$ yield the Schrödinger and Heisenberg pictures respectively. In the Heisenberg picture, unitary evolution of the observable is related to $\hat{H}$ by the Heisenberg equation,

$$i\frac{d\hat{A}}{dt} = \left[\hat{A}, \hat{H}\right]. \quad (1.20)$$

1.3.4 No-Cloning and Heisenberg Uncertainty

There exist a number of operations which are trivial to perform for classical systems, but which are not allowed for quantum states. For example, perfect duplication of an arbitrary unknown quantum state is impossible. To see this, consider a cloning machine $\hat{U}$ which copies an unknown state $|\psi\rangle$ onto an ancilla system, initially prepared in $|a\rangle$:

$$\hat{U}|\psi\rangle \otimes |a\rangle = |\psi\rangle \otimes |\psi\rangle. \quad (1.21)$$

If we use the machine to copy two particular quantum states, $|\psi\rangle$ and $|\varphi\rangle$, we have

$$\hat{U}|\psi\rangle \otimes |a\rangle = |\psi\rangle \otimes |\psi\rangle; \quad \hat{U}|\varphi\rangle \otimes |a\rangle = |\varphi\rangle \otimes |\varphi\rangle. \quad (1.22)$$

Taking the inner product of these two equations, we have $\langle\psi|\varphi\rangle = (\langle\psi|\varphi\rangle)^2$, immediately implying that such a cloning machine cannot be universal. Note that this does not preclude the preparation of an ensemble of identical states by repeated application of a trusted state preparation procedure.

Quantum mechanics also places fundamental limits on the extent to which the properties of a given ensemble of quantum states can be measured and known. Heisenberg's *uncertainty principle* states that: given as a resource an ensemble of identical unknown states $|\psi\rangle$, the standard deviation $\Delta(\hat{C})$, $\Delta(\hat{D})$ in measurements of two observables $\hat{C}$, $\hat{D}$ is bounded below by

$$\Delta(\hat{C})\Delta(\hat{D}) \geq |\langle\psi| \left[\hat{C}, \hat{D}\right] |\psi\rangle|/2. \tag{1.23}$$

That is, when $\hat{C}$ and $\hat{D}$ do not commute, the better our knowledge of C, the less information we have on D. The related (but distinct) principle of *complementarity* further limits our ability to measure noncommuting observables of quantum states, and is described in Sect. 3.2.2.

1.3.5 Qubits

The basic unit of classical information is the *bit*, $b \in \{0, 1\}$. The quantum analogue is the *qubit*, a two-level quantum system with Hilbert space dimension $d = 2$. By analogy with classical bits, the states $|0\rangle$, $|1\rangle$ form a basis for $\mathscr{H}$, and a single qubit can occupy any normalized superposition state

$$|\psi\rangle = \alpha|0\rangle + \beta|1\rangle; \quad |\alpha|^2 + |\beta|^2 = 1. \tag{1.24}$$

Neglecting a global phase, this can be re-written as

$$|\psi\rangle = \cos\frac{\theta}{2}|0\rangle + e^{i\varphi}\sin\frac{\theta}{2}|1\rangle, \tag{1.25}$$

leading to a natural geometrical representation of the state space $\mathscr{H}$ of the qubit as the surface of a unit sphere, often referred to as the *Bloch sphere* (Fig. 1.2). Throughout this thesis we will make use of the quadrant points of the Bloch sphere

$$|0\rangle \equiv \begin{bmatrix}1\\0\end{bmatrix}, \quad |+\rangle \equiv \tfrac{1}{\sqrt{2}}\left(|0\rangle + |1\rangle\right), \; |+i\rangle \equiv \tfrac{1}{\sqrt{2}}\left(|0\rangle + i|1\rangle\right) \tag{1.26}$$

$$|1\rangle \equiv \begin{bmatrix}0\\1\end{bmatrix}, \quad |-\rangle \equiv \tfrac{1}{\sqrt{2}}\left(|0\rangle - |1\rangle\right), \; |-i\rangle \equiv \tfrac{1}{\sqrt{2}}\left(|0\rangle - i|1\rangle\right), \tag{1.27}$$

which are eigenstates of the Pauli matrices $\hat{\sigma}_z$, $\hat{\sigma}_x$, $\hat{\sigma}_y$ respectively.

Almost any two-level quantum system can be used to encode a qubit. Specific conditions for a qubit to be useful for quantum computation are given in Sect. 1.4.1. Qubit encodings for linear optics are discussed in Sects. 1.6.1 and 2.2.5.

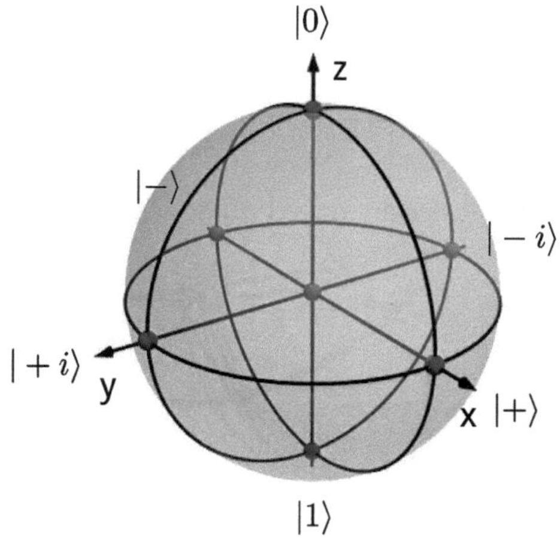

Fig. 1.2 The Bloch sphere provides a geometrical representation of the state space of a two-level quantum system—a qubit. Points on the surface of the sphere are pure ($|\psi\rangle = \alpha|0\rangle + \beta|1\rangle$, $\alpha^2 + \beta^2 = 1$), and include the quadrant points $|0\rangle$, $|1\rangle$, $|+\rangle$, $|-\rangle$, $|+i\rangle$, $|-i\rangle$. These points are eigenstates of the Pauli matrices $\hat{\sigma}_x$, $\hat{\sigma}_y$, $\hat{\sigma}_z$, and the axes are labelled correspondingly. The point at the centre of the sphere is the maximally mixed state, **1**

1.3.6 Mixture

So far we have only been concerned with closed quantum systems, where there is no uncontrolled outside influence, and all components of the system are accounted for. In practice, we often encounter situations in which some part of the quantum system is inaccessible to the experimentalist, often due to coupling to the environment. Under such circumstances, many of the assumptions of the previous discussion do not hold: namely, time evolution is no longer necessarily unitary, measurements are not guaranteed to be orthogonal projectors, and it is no longer satisfactory to represent states as rays in $\mathscr{H}$.

In order to represent the state of a quantum system subject to unknown external influence, we can consider a black-box device. We send into this device a quantum state, for example $|0\rangle$. Inside the box, a demon flips a fair coin. Depending on the outcome of the coin flip, the demon then outputs either the state $|0\rangle$, or the state $|1\rangle$. Now, we should not write the state of the ensemble generated by this box as a coherent superposition $|+\rangle = \frac{1}{\sqrt{2}}(|0\rangle + |1\rangle)$, as the two states are chosen according to a classical probabilistic process. We instead describe the state using a *density matrix* $\hat{\rho}$, defined as

$$\hat{\rho} \equiv \sum_i p_i |\psi_i\rangle\langle\psi_i|. \tag{1.28}$$

For the simple example cited here, the output of the box can be written as

$$\hat{\rho} = \frac{1}{2}|0\rangle\langle 0| + \frac{1}{2}|1\rangle\langle 1| = \frac{1}{2}\begin{bmatrix} 1 & 0 \\ 0 & 1 \end{bmatrix} = \mathbf{1}/2, \tag{1.29}$$

which is in contrast with the density matrix of the superposition state $|+\rangle$

$$\hat{\rho}_{+} = |+\rangle\langle +| = \frac{1}{2}\begin{bmatrix} 1 & 1 \\ 1 & 1 \end{bmatrix}. \tag{1.30}$$

All physical density matrices are semidefinite positive ($\hat{\rho} \geq 0$), Hermitian ($\hat{\rho} = \hat{\rho}^{\dagger}$), and have trace one ($\mathrm{Tr}(\hat{\rho}) = 1$): Time evolution of a density matrix $\hat{\rho}$ by a unitary process $\hat{U}$ proceeds as

$$\hat{\rho} = \sum_{i} p_i |\psi_i\rangle\langle\psi_i| \xrightarrow{\hat{U}} \hat{\rho} = \sum_{i} p_i \hat{U}|\psi_i\rangle\langle\psi_i|\hat{U}^{\dagger} = \hat{U}\hat{\rho}\hat{U}^{\dagger}, \tag{1.31}$$

and the expectation value of an observable $\hat{A}$ given a state $\hat{\rho}$ is given by $\langle\hat{A}\rangle = \mathrm{Tr}(\hat{A}\hat{\rho})$. Density matrices provide the most general description of quantum states. In the limit of zero coherence, the language of density operators reproduces classical probability theory. A standard approach for the description and characterization of open quantum processes is given in Sect. 2.7, and a method for the generation of mixed states from entangled two-qubit states is discussed in Sect. 2.9.

1.3.6.1 Purity

Uncertainty in a discrete classical random variable X is captured by the Shannon entropy,

$$H(X) \equiv -\sum_{i} p(x_i) \log p(x_i). \tag{1.32}$$

$H(X) = 1$ when X is the output of a single toss of a fair coin, and $H(X) = 0$ when, for example, $X \in x_0, x_1$ and $p(x_0) = 1$, $p(x_1) = 0$.

Mixed states can be thought of as possessing greater uncertainty than pure states, since for a maximally mixed state there exists no measurement basis $\{|\tau\rangle\}$ in which measurement outcomes are deterministic. In order to quantify uncertainty for a quantum *state* we might try to apply the Shannon entropy to measurement outcomes—however, this does not give the desired behaviour. If $\hat{\rho}$ is a pure state $|\psi\rangle\langle\psi|$, then there is a conjugate measurement basis $\{|\psi\rangle^{*}, |\psi_{\perp}\rangle^{*}\}$ in which the measurement outcome *is* deterministic. If we assign eigenvalues ± 1 to each basis state respectively we will *always* register $+1$, giving a Shannon entropy over measurement outcomes of $H(X) = 0$. However, we could equally choose to measure in a diagonal basis, giving uniformly distributed random measurement outcomes and thus $H(X) = 1$.

So, we cannot use the Shannon entropy to quantify uncertainty for a particular state, as a good measure of states should be independent of any choice of measurement basis. The *von Neumann entropy* is an entropic measure which solves this problem. It is defined in a very similar way to the Shannon entropy: the von Neumann entropy of a state $\hat{\rho}$ with a spectral decomposition $\{\lambda_i\}$, $\{|\lambda_i\rangle\}$ is

$$S(\hat{\rho}) \equiv -Tr\left(\hat{\rho}\log\left(\hat{\rho}\right)\right) = -\sum_i \lambda_i \log\left(\lambda_i\right) = -\langle \log \hat{\rho} \rangle \tag{1.33}$$

which evaluates to 0 for all pure states, is maximal and equal to $\log d$ for all maximally mixed states (where d is the dimension of the Hilbert space), and increases monotonically with all sensible measures of mixture. Further useful measures of the degree of mixture of a quantum state are given by two related quantities, the *purity*

$$\gamma(\hat{\rho}) \equiv Tr(\hat{\rho}^2) \tag{1.34}$$

and the *linear entropy* $S_L(\hat{\rho}) \equiv 1 - \gamma(\hat{\rho}^2)$. The purity of a pure state $|\psi\rangle$ is $Tr(|\psi\rangle\langle\psi||\psi\rangle\langle\psi|) = 1$, and for a maximally mixed state $\gamma(\mathbf{1}) = 1/d$.

1.3.7 Entanglement

Superposition states of a single particle, permitted by quantum mechanics as previously described, have powerful and counterintuitive implications. Single-particle experiments such as Young's double slit (Sect. 3.2) show qualitative differences in physical behaviour with respect to classical mechanics, and quantum algorithms such as Grover search (Sect. 1.4.1) can provide a polynomial speedup for certain computational tasks.

However, in order to fully appreciate the extent to which quantum mechanics is profoundly distinct from classical physics, it is important to consider multi-particle experiments involving the related phenomena of *entanglement* and *nonlocality*. Using these phenomena we can construct games which can provably only be be won by quantum players, and experimentally falsify the extremely natural and widely-held notion of a local-realistic universe. Entanglement is the *resource* which drives most quantum technologies, including quantum computing, metrology, simulation, and some schemes for quantum communication. Throughout this thesis, we make use of entangled quantum states both as a resource for computation (Chap. 5 and Sect. 6.3.2) and as a basic physical phenomenon of fundamental interest (Chaps. 3, 4, and Sect. 6.3).

Einstein, whose celebrated theory of relativity restored locality to macroscopic physics, was intimately involved in the discovery [6], along with Podolsky, Rosen,

Schrödinger, and von Neumann, that quantum mechanics permits multipartite systems to exist in states which cannot be written as a product of their subsytems, i.e.

$$\hat{\rho}_{A,B,C\ldots} \neq \sum_i p_i \hat{\rho}_A \otimes \hat{\rho}_B \otimes \hat{\rho}_C \cdots \tag{1.35}$$

Quantum states which cannot be written in this form are said to be *entangled*. For such states, full knowledge of the individual subsystems does not imply full knowledge of the true, holistic state, and vice-versa. To see the physical effect of this phenomenon, we can consider the example of a bipartite entangled state of two qubits, shared between distant parties, Alice and Bob:

$$|\Phi^+\rangle = \frac{1}{\sqrt{2}}\left(|0_A 0_B\rangle + |1_A 1_B\rangle\right). \tag{1.36}$$

This state cannot be written as the product of two separate objects, as in (1.35). When both parties measure their system in the $\{|0\rangle, |1\rangle\}$ (logical) basis, we see that Alice and Bob each have 50 % probability of detecting 0 or 1, and their measurement outcomes are also strongly *correlated*—Alice's outcome is always the same as Bob's.

$$P_{00} = |\langle 00|\Phi^+\rangle|^2 = \frac{1}{2}; \quad P_{11} = |\langle 11|\Phi^+\rangle|^2 = \frac{1}{2}; \quad P_{01} = P_{10} = 0 \tag{1.37}$$

Correlated, probabilistic behaviour indistinguishable from that generated by this state *when measuring in the logical basis* can easily be simulated classically. Flip a coin, and if it outputs heads, give to Alice and Bob the state $|00\rangle$, otherwise provide $|11\rangle$, i.e. generate the mixed state $(|00\rangle\langle 00| + |11\rangle\langle 11|)/2 = \mathbf{1}$. The troubling observation that led Einstein, Podolsky and Rosen (EPR) to conclude that quantum mechanics was "incomplete" becomes apparent when Alice measures in an arbitrary basis $\{|\lambda_0\rangle, |\lambda_1\rangle\}$.

Depending on Alice's measurement outcome, she will remotely project Bob's state onto one of the conjugate basis states $\{|\lambda_0\rangle^*, |\lambda_1\rangle^*\}$, leaving the entire system in $|\lambda_{0_A}\lambda_{0_B}\rangle^*$ or $|\lambda_{1_A}\lambda_{1_B}\rangle^*$ (see Sect. 1.3.2). The implication of this effect, named *steering* by Schrödinger, is that either (i) the physical state of Bob's particle was somehow remotely and instantaneously modified by Alice's choice, or (ii) Bob's state was never well-defined in the first place. Put another way, either the universe is *nonlocal*—meaning that the relationship between two separate objects cannot be completely accounted for by a set of factors that previously acted on those objects—or it is not *realistic*—the physical properties of objects do not have real, pre-existing values, until a measurement is made—or both.

Entanglement can be measured in myriad different ways, and a full discussion of the diverse variety of entanglement measures and associated partitionings of Hilbert space is beyond the scope of this thesis. A comprehensive review was given by Plenio [7]. We provide here a minimal set of examples, as reference points which will be used throughout this thesis.

We can assert some simple and reasonable conditions for a measure of entanglement:

- Separable states of the form (1.35) contain no entanglement.
- Entanglement cannot be increased through local operations and classical communication (LOCC) alone. Experimentalists in separate labs, connected only by classical channels and each having access to one subsystem of a larger quantum state, cannot increase the extent to which they are entangled.[1] This implies that *entanglement is invariant under local unitaries*. A state $\hat{\rho}$ can be said to be at least as entangled than another $\hat{\rho}'$ if $\hat{\rho}$ can be converted to $\hat{\rho}'$ through LOCC operations alone.
- Maximally entangled states exist. The *Bell states*

$$|\Psi^{\pm}\rangle \equiv \frac{1}{\sqrt{2}}\left(|01\rangle \pm |10\rangle\right); \qquad |\Phi^{\pm}\rangle \equiv \frac{1}{\sqrt{2}}\left(|00\rangle \pm |11\rangle\right) \tag{1.38}$$

form an orthonormal basis set for two-qubit states, and are the canonical example of two-qubit maximally entangled states. Any pure or mixed state of two qubits can be prepared from a Bell state using only LOCC operations, and one can easily convert between Bell states using only local unitary operations $\hat{U}_A \otimes \hat{U}_B$. For multipartite systems, a satisfactory definition of maximally entangled states has proved elusive—see, for example, results by Greenberger et al. [8].

Two entanglement measures of particular relevance to experimental quantum optics are the *entropy of entanglement* and the *concurrence*. As we have already seen, individual subsystems of an entangled state are strongly dependent on one another. If Alice and Bob share the separable pure state $\hat{\rho}_{AB} = |0_A 0_B\rangle\langle 0_A 0_B|$, the reduced density matrix of Alice, tracing over Bob's state, is $\hat{\rho}_A = \mathrm{Tr}_B \hat{\rho}_{AB} = |0\rangle\langle 0|$—that is, her state is pure and independent of Bob's system. However, when Alice and Bob share a maximally entangled state (for example $|\Phi^+\rangle\langle\Phi^+|$), although the state of the whole system is pure, Alice's reduced density matrix is maximally mixed, $\hat{\rho}_A = \frac{1}{2}\left(|0\rangle\langle 0| + |1\rangle\langle 1|\right) = \mathbf{1}$. We can use this behaviour to devise an entanglement measure for pure states based on the generalized quantum uncertainty of the state of one subsystem, tracing over the other, where uncertainty is characterized by the von Neumann entropy.

The *entropy of entanglement* is defined [9] as

$$E(\hat{\rho}_{AB}) = S(\hat{\rho}_B) = S(\hat{\rho}_A) = S\left[Tr_B\left(\hat{\rho}_{AB}\right)\right]. \tag{1.39}$$

In this example of two qubits, it is natural to choose a base-2 logarithm, in which case E ranges from zero, when $\hat{\rho}_{AB}$ is separable, to $\log_2 d = 1$ for a maximally entangled two-qubit state. A nice property of $E(\hat{\rho}_{AB})$ is that for two qubits, in the asymptotic

[1] Experimentalists *can* use LOCC operations to selectively *throw away* states coming from some partially entangled source, thus producing a postselected state with greater entanglement than the source itself. However, the entanglement of the system as a whole, including those systems that were thrown away, does not increase under LOCC operations.

limit of many experiments, E is equal to the ratio m/n, where m is the number of perfect, maximally entangled singlet states that can be reversibly generated by LOCC operations from a source producing n copies of $\hat{\rho}_{AB}$.

The entropy of entanglement is defined only for pure states. A useful entanglement measure which also works for mixed states is the *concurrence*, defined for a mixed state of two qubits $\hat{\rho}$ as

$$\mathcal{C}(\hat{\rho}) \equiv \max(0, \lambda_1, \lambda_2, \lambda_3, \lambda_4) \tag{1.40}$$

where $\{\lambda_i\}$ are eigenvalues of the matrix $R = \sqrt{\sqrt{\hat{\rho}}\tilde{\rho}\sqrt{\hat{\rho}}}$ and $\tilde{\hat{\rho}} = (\hat{\sigma}_y \otimes \hat{\sigma}_y)\hat{\rho}(\hat{\sigma}_y \otimes \hat{\sigma}_y)$. $\mathcal{C}$ ranges from 0 for a separable state and 1 for a maximally entangled state, and is monotonically related to E. We make use of the concurrence in Sect. 4.5 and Chap. 5 of this thesis.

1.3.8 Bell Nonlocality

In our discussion so far, it has been necessary to use the formalism of state vectors, operators, measurements and so on in order to provide an intuitive picture of the character and effects of entanglement. Although we will see later on that machines which use entanglement as a resource have the potential to dramatically affect the real classical world, it is hard to give a good picture of the *fundamental properties* of entanglement without appealing to the quantum mechanical formalism. However, it turns out that we can construct experiments which reveal—without *fully* characterising entanglement itself—the sharp separation between allowed behaviour of entangled *vs* separable states, without the need to first choose an in-depth physical model of the world, and which rely only on simple statements about space and probability.

Classical physics is local. Consider two parties, Alice and Bob, who are separated in space by many light-years. They each possess (Fig. 1.3) a single object. Their objects may have originated from a common source. Alice and Bob now independently and freely choose to measure their respective objects in some way. We do not need to use the quantum mechanical description of measurement—we simply imagine switches allowing Alice to measure in $a \in \{a_0, a_1 \ldots\}$ and Bob in $b \in \{b_0, b_1 \ldots\}$, yielding measurement outcomes A and B respectively. When this experiment is repeated many times, these measurement outcomes are governed by a probability distribution $p(AB|ab)$.

Alice and Bob's systems may have met in space at some point in their history, and may have been prepared or choreographed in a particular way, giving rise to correlations or dependencies in p. We denote this prior knowledge by a *local (hidden) variable* λ, which accounts for any local information or "hidden pre-programming" which these objects might possess. Having done so, we define a *local theory* as

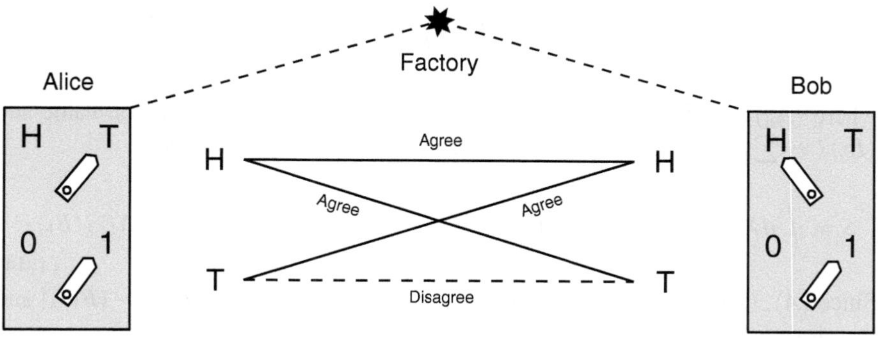

Fig. 1.3 A Bell-CHSH test. *Alice* and *Bob* receive devices from a common source or factory. Each device has a binary input (heads, H or tails, T) and a binary output (0, 1). The internal machinery of the devices, as well as the pre-arranged strategy of *Alice* and *Bob*, are left unspecified—the only condition is that the devices are separated in space and cannot communicate. Having received their devices, *Alice* and *Bob* each flip a coin, obtaining H or T. Their task is then to satisfy the rules illustrated in the central schematic. Namely, if one or more coins shows heads, the output of *Alice* and *Bob's* devices should *Agree*, yielding $0_a 0_b$ or $1_a 1_b$. Only when both parties flip tails should they *Disagree*, outputting $0_a 1_b$ or $1_a 0_b$. It is easily confirmed that all local strategies are limited to a probability of success of $3/4$. However, when *Alice* and *Bob* share an entangled state, this bound can be violated

one in which we can factorize the probability distribution [10] over measurement outcomes as

$$p(AB|ab) = \int_\Lambda d\lambda \, q(\lambda) p(A|a, \lambda) p(B|b, \lambda), \tag{1.41}$$

where q is a random variable over all possible $\lambda \in \Lambda$, which takes into account the possibility that λ may change between measurement runs. The outcome of Alice's measurement thus does not depend on Bob's choice of measurement operator, and is fully described by local effects. Note that we arrive at this definition without any particular choice of physical model.

In 1964, John Bell proved [11] that the predictions of quantum theory are incompatible with the notion of locality captured in (1.41). Since 1964, many variations on Bell's proof have been developed, some of which are simpler to derive, or experimentally test, than others. Here we consider a Bell test due to Clauser et al. [12], in which we assume only two measurement settings $a \in \{a_0, a_1\}$, $b \in \{b_0, b_1\}$, and two measurement outcomes $A, B \in \{-1, +1\}$ per party.

Consider the quantity

$$S = \langle A_0 B_0 \rangle + \langle A_0 B_1 \rangle + \langle A_1 B_0 \rangle - \langle A_1 B_1 \rangle \tag{1.42}$$

where $\langle A_a B_b \rangle = \sum_{A,B} ABp(AB|ab)$ is the expectation value of the product $A \cdot B$, given measurement settings a, b. Assuming a local model, we re-write these expectation values using (1.41),

$$\langle A_a B_b \rangle = \int d\lambda q(\lambda) \langle A_a \rangle_\lambda \langle B_b \rangle_\lambda \tag{1.43}$$

where $\langle A_a \rangle_\lambda = \sum_A Ap(A|a, \lambda) \in [-1, 1]$ is Alice's local expectation value and $\langle B_b \rangle_\lambda = \sum_B Bp(B|b, \lambda) \in [-1, 1]$ is Bob's. Now, (1.42) becomes

$$S = \int d\lambda S_\lambda = \int d\lambda \langle A_0 \rangle_\lambda \langle B_0 \rangle_\lambda + \langle A_0 \rangle_\lambda \langle B_1 \rangle_\lambda + \langle A_1 \rangle_\lambda \langle B_0 \rangle_\lambda - \langle A_1 \rangle_\lambda \langle B_1 \rangle_\lambda. \tag{1.44}$$

Since $\langle A \rangle, \langle B \rangle \in [-1, 1]$, we see that $|S| \leq |\langle B_0 \rangle_\lambda + \langle B_1 \rangle_\lambda + \langle B_0 \rangle_\lambda - \langle B_1 \rangle_\lambda|$ and therefore

$$|S| \leq 2. \tag{1.45}$$

This is the Bell-CHSH inequality, which holds for all local realistic models. Now consider a scenario in which Alice and Bob share the Bell state $|\psi_{AB}\rangle = |\Psi^-\rangle$. Their local measurement settings are now described by qubit measurement operators, $\hat{A}_i$, $\hat{B}_i$. It is helpful to express each measurement operator as a *Bloch vector*, which maps a single-qubit measurement operator to $\mathbb{R}^3$.

$$\hat{A}_i = a_{i_1}\hat{\sigma}_x + a_{i_2}\hat{\sigma}_y + a_{i_3}\hat{\sigma}_z = \vec{a}_i \cdot \vec{\sigma}, \tag{1.46}$$

$$\hat{B}_i = b_{i_1}\hat{\sigma}_x + b_{i_2}\hat{\sigma}_y + b_{i_3}\hat{\sigma}_z = \vec{b}_i \cdot \vec{\sigma}. \tag{1.47}$$

Single-qubit measurement operators can thus be visualized in the Bloch sphere (Fig. 1.2). Setting $q(\lambda) = 1$, it is then easy to show that the expectation value of AB is simply related to the overlap of $\vec{a}$ and $\vec{b}$,

$$\langle \hat{A}_i \hat{B}_i \rangle_\psi = \int d\lambda q(\lambda) \langle \psi | \hat{A}_i \otimes \hat{B}_i | \psi \rangle = -\vec{a}_i \cdot \vec{b}_i = -\cos(\theta), \tag{1.48}$$

where θ is the angle between $\vec{a}_i$ and $\vec{b}_i$. If they choose the following measurement operators

$$\hat{A}_0 = \hat{\sigma}_z; \quad \hat{A}_1 = \hat{\sigma}_x; \quad \hat{B}_0 = -\frac{\hat{\sigma}_z + \hat{\sigma}_x}{\sqrt{2}}; \quad \hat{B}_1 = \frac{\hat{\sigma}_z - \hat{\sigma}_x}{\sqrt{2}} \tag{1.49}$$

it is easy to show that $S = 2\sqrt{2} > 2$, violating (1.45). The maximum value of S which can be obtained, using any quantum state, is $2\sqrt{2}$. Moreover, this value is only obtained for maximally entangled states. Considerable insight into the fundamental nature of quantum mechanics has been gained [10, 13, 14] through the construction of unphysical models or objects which violate Clauser-Horne-Shimony-Holt (CHSH) *beyond* $2\sqrt{2}$.

Since the discovery of Bell's theorem and its later development by CHSH, this inequality has been experimentally violated many times. Arguably the first robust experimental demonstration was made in 1982 by Aspect et al. [15], using entangled photon pairs from a calcium cascade source. More recent experimental implementations have focussed either on closing the *loopholes* which leave such experiments open to local-realistic interpretation [16, 17], or on the potential communication applications of nonlocal correlations, in the form of *device independent quantum key distribution* (see Sect. 1.4.2).

1.3.8.1 Obtaining Nonlocality

Not all entangled states exhibit nonlocal statistics. Pure states with any nonzero value of the entropy of entanglement (1.39), for instance states of the form

$$\sqrt{1-p}|\Psi^-\rangle + \sqrt(p)|00\rangle \tag{1.50}$$

violate Bell-CHSH (although not maximally) for all $p > 0$. In contrast, Werner [18] described mixed, entangled two-qubit states, showing EPR correlations and which cannot be written as (1.35), which do *not* exhibit nonlocal correlations. The *Werner state* with visibility V

$$\hat{\rho}_V \equiv V|\Psi^-\rangle\langle\Psi^-| + (1-V)\frac{\mathbb{1}}{4} \tag{1.51}$$

cannot violate Bell-CHSH for $V < 1/\sqrt{2}$.

Even if Alice and Bob share a maximally entangled state, nonlocal statistics are not always revealed—for instance, if they choose their measurements from a single orthogonal basis set. Therefore in order for Alice and Bob to guarantee that they will see nonlocal correlations, they must somehow co-ordinate their choice of measurement settings. This point is discussed in detail in Chap. 4 of this thesis.

1.4 Quantum Technologies

Information is physical.
Rolf Landauer

Information must necessarily be encoded in the state a physical system. In order to encode classical information, almost any physical system will do: human beings have cut giant figures into the chalky substrate of the Chiltern hills, hewn laws into stone tablets, and currently store exabytes of data in the magnetic domains of hard disk drives. Over the past century, with the advent of quantum mechanics, it came to be understood that information stored in the state of a quantum system—*quantum information*—is very distinct from its classical counterpart.

Quantum information is encoded in the probability amplitudes of a quantum state, and can therefore exist in an arbitrary coherent superposition. It follows that quantum information can be encoded in an entangled state, and can thus exhibit correlations which are classically forbidden. Moreover, as has already been discussed, the fact that quantum states cannot be cloned places restrictions on the extent to which quantum information can be reliably "read out" in a single shot.

These fundamental differences between allowed representations and operations on classical and quantum information lead to new applications, devices, and technologies, which cannot be accomplished by classical means. Specifically, quantum information science has revealed fundamentally new modes of information processing, measurement, communication, and simulation, which we detail below.

1.4.1 Quantum Computing

Quantum systems exhibit classically forbidden phenomena. As a result, quantum information can be processed using operations which are forbidden for classical machines. In particular, unitary evolution of quantum information can lead to *interference* effects which do not occur under stochastic evolution of classical information. This leads to the possibility of a *quantum computer*: an entangled, quantum, problem-solving machine. It has been shown that by exploiting these new operations, a quantum computer could in principle solve certain computational tasks using exponentially fewer resources than any classical machine.

To see that quantum information can be advantageously processed using classically forbidden operations, we look to a simple and concrete example due to Deutsch and Jozsa [19]. Given an unknown Boolean function f, the task is to determine whether f is *constant*, $f(0) = f(1)$, or *balanced*, $f(0) \neq f(1)$ (i.e. we want the parity of f). To answer this question classically, we must make two calls to f:

$$f(0) \oplus f(1) = \begin{cases} 0 \text{ if } f \text{ is constant} \\ 1 \text{ if } f \text{ is balanced} \end{cases} \tag{1.52}$$

where $\oplus$ denotes addition mod 2. However, implementing f using a two-qubit entangling gate $\hat{U}_f|x\rangle|a\rangle = |x\rangle|f(x) \oplus a\rangle$, we can effectively make a single call to f with a superposition of both arguments at once

$$\hat{U}_f\,|+\rangle \otimes |-\rangle = |0\rangle \otimes |f(0) \oplus 0\rangle - |0\rangle \otimes |f(0) \oplus 1\rangle \tag{1.53}$$

$$+\,|1\rangle \otimes |f(1) \oplus 0\rangle - |1\rangle \otimes |f(1) \oplus 1\rangle. \tag{1.54}$$

Applying a Hadamard operation to the first qubit, complex amplitudes in (1.54) destructively interfere to give

$$(\hat{H} \otimes \mathbf{1})\hat{U}_f|+\rangle \otimes |-\rangle = \begin{cases} \pm|0\rangle \otimes |-\rangle \text{ if } f \text{ is constant} \\ \pm|1\rangle \otimes |-\rangle \text{ if } f \text{ is balanced} \end{cases} \tag{1.55}$$

Measurement of the first qubit in the logical basis then immediately reveals the nature of f. Note that we only obtain a *global* property of f, not full information on the mapping (see Sect. 6.3.5). This algorithm is easily generalized to systems of n qubits, where it requires exponentially fewer calls to f with respect to all classical algorithms.

The Deutsch-Josza algorithm provides an attractive illustration of the characteristic properties of many quantum algorithms—dependence on interference of complex amplitudes, qubits, entangling gates, and ultimately an exponential speedup over classical machines. Unfortunately, the *problem* of Deutsch-Josza is rather contrived, and this algorithm has no known useful application.[2] Moreover, at the cost of deterministic operation, randomized classical algorithms perform very well at this task, classifying f in polynomial time, and furthermore the leap from f to the oracular $\hat{U}_f$ arguably renders the quantum-classical comparison somewhat unrealistic. As a result, the main utility of Deutsch-Josza is pedagogical. A similar role is played by *Grover search*, an algorithm first described in 1995 by Lov Grover. Grover's algorithm uses a single quantum system, together with a specific class of oracle, to accomplish a polynomial speedup over classical machines for a task resembling database search.

Long before Deutsch-Josza and Grover search, Feynman [20, 21] laid out the first strong argument as to why one might build a quantum computer. Feynman argued that since the state of a quantum system can exist in a coherent superposition over all allowed eigenstates, and since a system of n particles has exponentially many eigenstates in general, it is likely exponentially hard to simulate such systems using a classical computer. Feynman went on to propose that a quantum computer or *quantum simulator* should be capable of reproducing the dynamics of a system of interest, in a controlled way, using only polynomial resources. We can imagine that much as aircraft wings are numerically simulated prior to construction, drugs, materials and other atomic-scale systems might be designed on a quantum computer prior to synthesis in the laboratory. This application is potentially economically very significant, and would have a dramatic effect on science, medicine, and engineering. Quantum simulation is discussed in further detail in Chap. 5 and Sect. 6.3 of this thesis.

Quantum simulation has almost the opposite problem to Grover search and Deutsch-Josza. Quantum simulators constitute arguably the most practically useful known application of a quantum computer, but it remains very hard to prove either (i) that atomic/molecular systems of interest cannot be efficiently simulated by a classical machine or (ii) that all physical systems *can* be efficiently simulated by a quantum computer!

In 1994, Peter Shor first described an algorithm [22] which has since become the best-known proposed application for quantum computation. Shor showed that a uni-

[2] In terms of computational complexity, Deutsch-Josza provides an oracle relative to which EQP (the class of problems exactly soluble by a quantum computer in polynomial time) is distinguishable from P (decision problems soluble in poly-time by a deterministic Turing machine). However, we do not expect that a Deutsch-Josza machine would have direct "economically significant" implications!

versal quantum computer, capable of manipulating, entangling and measuring a large number of qubits, could be used to solve the *prime factoring* and *discrete logarithm* problems in polynomial time. This was an extremely powerful result, as the problem of prime factoring is strongly believed to be computationally intractable for classical machines, and is also useful for real-world practical tasks. Specifically, prime factoring is the task of identifying the prime factors a, b of a (large) composite L-bit number $N = ab$. The best-known classical algorithms run in time exponential in L, while Shor's algorithm runs in $O(L^3)$ time. A scalable implementation of Shor's factoring algorithm would break most (but not all) existing classical encryption algorithms, including Rivest-Shamir-Adleman (RSA) and elliptic-curve cryptography.

We have outlined above a few quantum algorithms most relevant to this discussion. A great many quantum algorithms have since been developed, most of which are outside the scope of this thesis. In Chap. 5, we introduce a new algorithm for simulation of quantum chemistry. In Sect. 6.3.2, we experimentally demonstrate a relatively new quantum algorithm, BosonSampling, which has particular relevance for the photonic platform addressed here.

1.4.1.1 The DiVincenzo Criteria

Although the quantum algorithms described above could in principle be implemented using special-purpose machines, one of the principal goals of quantum information science is the design and construction of *general-purpose*, universal quantum computers. Such a machine could be reconfigured, or *programmed*, to implement any conceivable quantum algorithm, and is arguably the most ambitious and potentially rewarding goal of the entire field of quantum information. The fact that a universal quantum computer could in principle be constructed under the known laws of quantum mechanics has been proven in works by Barenco, Bennett, Cleve, Deutsch, Ekert, DiVincenzo, Lloyd, Shor, Smolin, and many others. See, for example, Refs. [23–26].

In order to build such a machine we must first select a physical architecture, amenable to experimental implementation, in which to encode, manipulate and measure quantum information. Although we can construct quantum algorithms which provably cannot be efficiently performed by *any* known machine [1], any successful platform for quantum computing will require experimental resources which grow at most polynomially with the size of the quantum circuit, or the number of elementary operations required. In order to evaluate the suitability of proposed architectures and technologies for quantum computing, we make use of the *DiVincenzo criteria* [27]—the basic experimental criteria for any scalable platform for quantum computing. Here we list the five criteria most pertinent to our discussion:

- **A scalable system with well-characterized qubits** Single qubits, supporting coherent quantum superposition states, upon which quantum information can be encoded. A single qubit should not be prohibitively experimentally demanding to implement, and experimental resources should scale at most polynomially with the total number of qubits.

- **The ability to prepare a simple fiducial state** We must be sure of the initial state of the system. This fiducial state need not be entangled.
- **Evolution under a universal set of quantum gates** Lloyd [25], DiVincenzo [23] and many others have described small, discrete sets of elementary operations on qubits, which can be combined to implement any quantum algorithm. One example of such a *universal gate set* is formed by the (maximally entangling) two-qubit controlled-not (CNOT) gate, together with generic single-qubit operations. All such universal gate sets include at least one entangling operation.
- **Decoherence times much longer than the gate operation time** As has already been discussed (Sect. 1.3.6), interaction with the environment leads drives the state of the system towards a mixed state, in a process referred to as *decoherence*. Since a maximally-mixed state can be modelled by a classical probability distribution, decoherence almost always leads to failure of quantum algorithms. The characteristic rate at which the purity of the qubit state degrades, which is related to the strength of coupling to the environment, must therefore be slow with respect to the time taken to perform a gate operation.
- **Qubit measurement** The architecture must allow single-qubit quantum measurements, as described in Sect. 1.3.2. Measurement in the z-basis can be combined with a universal gate set to evaluate any possible observable on the system of qubits.

Quantum computation is widely believed to be the most technically challenging of all proposed quantum technologies, and it is likely that any platform satisfying the DiVincenzo criteria would also be capable of implementing other applications, described in Sects. 1.4.2 and 1.4.3.

1.4.1.2 Fault Tolerance

No useful machine exists in a vacuum. All practical machines are subject to the influence of noise, error, and loss, due to both interaction with the environment and imperfect fabrication or operation of the machine itself. Classical computers overcome noise by two complementary methods. First, the reliability of individual components in modern classical computers is extremely good: typical error rates are on the order of 1 in 10×10^{17} operations. The overwhelming majority of these few errors are then detected and corrected by means of redundancy-based error-correcting codes. The simplest example is to encode the bit state 0 on n bits, 0000..., and similarly for 1, in which case error can be exponentially suppressed by means of a simple majority-voting system.

Error correction is similarly essential for quantum computers. Without it, the probability of success of any realistic quantum computation falls off exponentially as the system evolves in time. Fortunately, a number quantum ECCs [1, 28–30] have been developed which effectively protect quantum states against noise. Owing in part to the no-cloning theorem, these codes are necessarily distinct from the simplest classical techniques, however they are still largely based on redundancy, in that a single *logical* qubit is represented by a number of *system* qubits, whose state is monitored and adjusted to correct errors. As with classical ECCs, quantum ECCs therefore demand an overhead, in terms of both qubit and gate count, with respect to the naïve implementation. In practise, this overhead can be extremely large [31]. The overhead for a given choice of ECCs is guaranteed to be polynomial in problem size only when the intrinsic *error rate* is below a certain threshold value. This is the *threshold theorem* [32], without which scalable quantum computing would likely not be a realistic prospect.

1.4.2 Quantum Communication

Prime factoring can be seen as a *one-way* function, which is hard to compute, but easy to check. The security of almost all digital communication is currently guaranteed by the difficulty of the forward problem, which is the basis of the RSA algorithm for public-key cryptography. RSA provides a method by which two parties can securely communicate, without having to first share a large one-time pad. While RSA has been enormously successful, it is by no means perfect. First, it is not known whether factoring is *fundamentally* classically intractable: at any time, an efficient classical factoring algorithm could suddenly be discovered, breaking the security of RSA. Secondly, an eavesdropper with access to a scalable quantum computer could use Shor's algorithm to silently decrypt and listen-in on this communication.

While quantum information science enables a realistic attack on RSA, it also provides a new technique for secure communication, based on quantum theory itself. In 1984, Bennett and Brassard [33] (BB84) described a method allowing two distant parties to communicate securely over an untrusted channel, using quantum states as the information carrier. This technique, together with its many derivatives, is referred to as quantum key distribution (QKD). The security of QKD is guaranteed by the axioms of quantum mechanics, in particular the no-cloning theorem (Sect. 1.3.4). In the event that an eavesdropper successfully reads private information from the channel, the state of the quantum system carrying that information is measurably disturbed, in which case the honest parties cease communication. In order to eavesdrop on a channel secured by QKD,[3] an attacker would need to discover physical effects which contradict no-cloning, which would be considerably more surprising than the discovery of a polynomial-time classical factoring algorithm.

At the time of writing, QKD is one of the few quantum technologies to have reached the market. This reflects the relative experimental accessibility of the task. All commercial QKD systems use photons as the information carrier, owing to the many advantages described in Sect. 1.6.1. Most QKD systems either time-bin or polarization encoding, carrying single photons or weak coherent pulses over optical fibre or in free-space.

---

[3] Assuming a perfect experimental implementation, see Ref. [34].

Recently, Lydersen et al. reported a functional attack on commercial QKD systems [34], which exploits details of the technical implementation to gain control over the measurement apparatus and steal information. Device-independent quantum key distribution DI-QKD [35], which necessarily depends on entanglement and nonlocal correlations, has been proposed as a solution to this class of attack. In Chap. 4, we introduce a number of theoretical and experimental techniques which may facilitate DI-QKD in real-world scenarios.

1.4.3 Quantum Metrology

We have argued that since computation is a physical process, quantum mechanics can be used to compute. Moreover, an advantage in computation can be gained by using a quantum machine. Similarly, *measurement* is physical. It turns out that by using a quantum apparatus to probe a system of interest, a number of tangible advantages can be gained with respect to classical methods [36, 37].

Classical measurements are fundamentally limited by what is known as the *shot noise*, or the *standard quantum limit*. Averaging over n measurements of a given observable A, by the central limit theorem the statistical uncertainty in the measured value of A scales as

$$\Delta A \propto 1/\sqrt{n}. \tag{1.56}$$

However, by probing the sample using entangled quantum states such as those described in Sect. 1.5.3, followed by quantum measurement of the resulting state, this uncertainty can in principle be reduced to a reciprocal scaling $\Delta A \propto 1/n$, violating the standard quantum limit. This method, known as *quantum metrology*, is particularly advantageous when the sample is extremely fragile and prone to damage by the measurement process itself, as it allows the same amount of information to be obtained using fewer discrete measurements.

We recently performed an experimental implementation [38] of a new scheme for loss-tolerant quantum metrology [39], which makes use of the photon counting capability developed in Sect. 6.2. Unfortunately, this work was not completed in time for inclusion in this thesis.

1.5 Light

Throughout this thesis we will examine the use of quantum states of light as a testbed for fundamental quantum mechanical phenomena, as well as the basic substrate upon which quantum-photonic technologies are built. We now lay out a theoretical framework to describe both classical and quantum states of light, in particular the quantization of the electromagnetic field, following the approach of Venkataram [40]. The following analysis is presented in Gaussian units.

1.5.1 Light as a Wave

Classical electromagnetic effects are governed by Maxwell's equations:

$$\nabla \times \mathbf{H} = \frac{1}{c}\left(\frac{\partial \mathbf{D}}{\partial t} + 4\pi \mathbf{J}_f\right) \tag{1.57}$$

$$\nabla \times \mathbf{E} = -\frac{1}{c}\frac{\partial \mathbf{B}}{\partial t} \tag{1.58}$$

$$\nabla \cdot \mathbf{B} = 0 \qquad \nabla \cdot \mathbf{D} = 4\pi \rho_f \tag{1.59}$$

where $\mathbf{H}$ is the magnetic field, $\mathbf{D} = \varepsilon \mathbf{E}$ is the electric flux density, $\mathbf{J}_f$ is the free current density, $\mathbf{E}$ is the electric field, and $\mathbf{B} = \mu \mathbf{H}$ is the magnetic flux density. ρ_f is the free charge density or charge per unit volume, and c is the speed of light. The dielectric permittivity ε and the magnetic permeability μ are related to the dielectric and magnetic susceptibilities χ_e, χ_m by

$$\varepsilon(\mathbf{E}) = \varepsilon_0 \left[1 + \chi_e(\mathbf{E})\right] \tag{1.60}$$

$$\mu(\mathbf{H}) = \mu_0 \left[1 + \chi_m(\mathbf{H})\right] \tag{1.61}$$

where ε_0 and μ_0 are the permittivity and permeability of the vacuum, respectively.

In the absence of charges ($\rho_f = 0$) and currents ($\mathbf{J}_f = 0$), Maxwell's equations reduce to

$$\nabla \times \mathbf{E} = -\frac{1}{c}\frac{\partial \mathbf{B}}{\partial t}; \qquad \nabla \times \mathbf{B} = +\frac{1}{c}\frac{\partial \mathbf{E}}{\partial t}; \qquad \nabla \cdot \mathbf{E} = 0; \qquad \nabla \cdot \mathbf{B} = 0. \tag{1.62}$$

For convenience we have taken $c = 1/\sqrt{\mu\varepsilon}$ to be the phase velocity of light in the medium. The *refractive index* n of the material

$$n = \sqrt{\frac{\mu\varepsilon}{\mu_0\varepsilon_0}} = \frac{c_0}{c} \tag{1.63}$$

relates c to the speed of light in the vacuum c_0. Taking the curl ($\nabla\times$) of the first two expressions in (1.62) we arrive at the electromagnetic wave equations

$$\nabla^2 \mathbf{E} = \frac{1}{c^2}\frac{\partial^2 \mathbf{E}}{\partial t^2}; \qquad \nabla^2 \mathbf{B} = \frac{1}{c^2}\frac{\partial^2 \mathbf{B}}{\partial t^2}. \tag{1.64}$$

The solutions $\mathbf{E}(\mathbf{r}, t)$ and $\mathbf{B}(\mathbf{r}, t)$ to these equations represent time-dependent electric and magnetic fields—light—propagating through the medium at $c \sim 3 \times 10^8$ m/s. These solutions are subject to the constraints that $\mathbf{B}$ and $\mathbf{E}$ should be perpendicular both to each other and the axis of propagation, and in phase, but may otherwise be very varied in form.

One solution to (1.64) for an inhomogeneous dielectric is a linearly polarized monochromatic field with wavelength λ,

$$\mathbf{E}(\mathbf{r},t) = \mathbf{A}(\mathbf{r})\, e^{i(\omega t - \phi(\mathbf{r}))} \tag{1.65}$$

where $\omega = 2\pi c/\lambda$ is the angular frequency and $\mathbf{A}$ is the amplitude vector which determines the polarization. When the medium is homogeneous, or in free space, an even simpler solution is given by a *plane wave* travelling in the $\hat{z}$ direction

$$\mathbf{E}(\mathbf{r},t) = \mathbf{A}e^{i(\omega t - kz)} \tag{1.66}$$

where $k = \omega/c$ is the wavenumber.

We will now consider a *single* eigenmode of the electromagnetic field with wave vector $\mathbf{k} = k\hat{k}$, where $\hat{k}$ is a unit vector in the direction of propagation. For a mode $\mathbf{k}$, solutions of (1.64) can be separated into a time-dependent complex function $\alpha_{\mathbf{k}}(t)$ and a spatial function $\mathbf{E}_0(\mathbf{r})$, where by convention the electric field $\mathbf{E}_{\mathbf{k}}$ is taken to be the real part of the product of $\alpha_{\mathbf{k}}$ and $\mathbf{E}_0$

$$\mathbf{E}_{\mathbf{k}}(\mathbf{r},t) \equiv \mathrm{Re}(\alpha_{\mathbf{k}}(t)\mathbf{E}_0(\mathbf{r})) = \alpha^*_{\mathbf{k}}(t)\mathbf{E}^*_0(\mathbf{r}) + \alpha_{\mathbf{k}}(t)\mathbf{E}_0(\mathbf{r}). \tag{1.67}$$

For $\mathbf{E}_{\mathbf{k}}$ and $\alpha_{\mathbf{k}}$ to be consistent with the wave equation (1.64), they must satisfy

$$\alpha_{\mathbf{k}}(t) = \alpha_{\mathbf{k}}(0)e^{ickt}; \qquad \nabla^2\mathbf{E}_{\mathbf{k}} + k^2\mathbf{E}_{\mathbf{k}} = 0. \tag{1.68}$$

The second of these two expressions is the Helmholtz equation. The magnetic field must be perpendicular to both $\mathbf{E}$ and the direction of propagation, $\mathbf{B}_{\mathbf{k}}\,(\mathbf{r},t) = \hat{k} \times \mathbf{E}_{\mathbf{k}}\,(\mathbf{r},t)$ and is thus related to $\alpha(t)$ and $\mathbf{E}_0(\mathbf{r})$ by

$$\mathbf{B}(\mathbf{r},t) = \frac{i}{k}\left[\alpha^*_{\mathbf{k}}(t)\nabla \times \mathbf{E}^*_0(\mathbf{r}) - \alpha_{\mathbf{k}}(t)\nabla \times \mathbf{E}_0(\mathbf{r})\right]. \tag{1.69}$$

In order to find the Hamiltonian of the electromagnetic field, we must integrate the energy density of the electric and magnetic fields over all space,

$$H = \int \mathcal{H} d^3 r = \int \frac{1}{8\pi}\left(\mathbf{E}^2 + \mathbf{B}^2\right) d\mathbf{r}. \tag{1.70}$$

Combining (1.67) and (1.69), together with careful choice of normalization of $\alpha(t)$ and $\mathbf{E}_0(\mathbf{r})$, we arrive at a Hamiltonian for the electromagnetic field in a mode $\mathbf{k}$, in terms of the ansatz $\alpha_{\mathbf{k}}(t)$

$$H_{\mathbf{k}} = \frac{\hbar c k}{2}\left(\alpha^*_{\mathbf{k}}\alpha_{\mathbf{k}} + \alpha_{\mathbf{k}}\alpha^*_{\mathbf{k}}\right) = \hbar c k|\alpha_{\mathbf{k}}|^2. \tag{1.71}$$

Here, $\hbar$ is simply a constant with units of action. As we will see in Sect. 1.5.2, this notation is chosen for a reason!

1.5.1.1 Interference

When two light fields occupy the same region of space, interference effects occur. The frequency of light is generally speaking too high (5×10^{14} Hz) for the electric field to be observed directly, and most measuring devices are only sensitive to the time-averaged intensity $I = \langle |\mathbf{E}(\mathbf{r}, t)|^2 \rangle$. The net electric field is the sum over modes, $\mathbf{E}(\mathbf{r}, t) = \sum_i \mathbf{E}_i(\mathbf{r}, t)$. For the simple example of interference of two linearly polarized monochromatic fields (1.65) $\mathbf{E}_1, \mathbf{E}_2$, the intensity observed at a point $\mathbf{r}$ is then given by

$$I(\mathbf{r}, t) = \langle |\mathbf{E}_1(\mathbf{r}, t)|^2 \rangle + \langle |\mathbf{E}_2(\mathbf{r}, t)|^2 \rangle + \langle \mathbf{E}_1 \cdot \mathbf{E}_2^* \rangle + \langle \mathbf{E}_1^* \cdot \mathbf{E}_2 \rangle \tag{1.72}$$

$$= I_1 + I_2 + 2(\mathbf{A}_1 \cdot \mathbf{A}_2) \cos\left[(\omega_1 - \omega_2)t - (\phi_1(\mathbf{r}) - \phi_2(\mathbf{r}))\right]. \tag{1.73}$$

We thus observe sinusoidal interference patterns in the measured intensity, depending on the relative phase and frequency of the two sources. Note that the strength or *contrast* of the observed interference fringe

$$C \equiv \frac{I_{\text{max}} - I_{\text{min}}}{I_{\text{max}} + I_{\text{min}}} \propto \mathbf{A}_1 \cdot \mathbf{A}_2 \tag{1.74}$$

depends on the polarization of the two sources: if they have orthogonal polarization the $(\mathbf{A}_1 \cdot \mathbf{A}_2)$ term vanishes and $C \to 0$.

1.5.1.2 Guided Modes

So far, our analysis has been focussed on light in a vacuum or homogeneous medium. Under these conditions, the propagation of laser light is well-approximated by Gaussian beam optics, in which the time-independent component of the electric field is normally distributed about the beam centre,

$$\mathbf{E}_0(\mathbf{r}) = \mathbf{E}_A \cdot e^{-||\mathbf{r}||^2/\omega_0^2}. \tag{1.75}$$

Throughout this thesis, as well as Gaussian beam optics in free space, we will make use of optical fibres and waveguides to confine and direct monochromatic light and single photons on-chip. These structures are constructed from two different materials: a *core* with refractive index n_1, in which the majority of the propagating electric field is confined, and a *cladding* constituting the substrate or surroundings of the waveguide, with index n_2. In this discussion we will describe the confinement and guiding of light in an idealized 1D rectangular waveguide, as shown in Fig. 1.4. All waveguides used in this thesis are rectangular. Further technical discussion of waveguide geometries and material systems is given in Sect. 1.6.5.

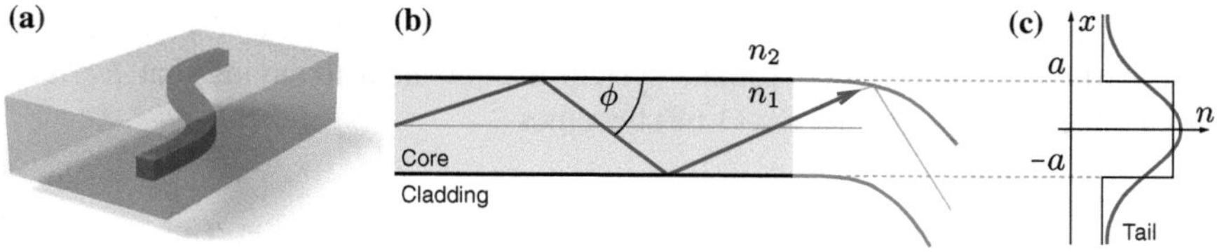

Fig. 1.4 Optical waveguides **a** Rectangular waveguide showing core and cladding, in a bend structure. **b** In the ray-optics picture, light is confined in the waveguide by total internal reflection when $n_1 \sin(\pi/2 - \phi) \geq n_2$. **c** 1D refractive index profile of a rectangular waveguide, and a single spatial mode

A working understanding of the confinement of light in an optical waveguide can be obtained from the ray-optics picture, in which a ray of light propagates in a straight line in the waveguide structure. At the interface between core and cladding, the ray is completely internally reflected if and only if the angle of incidence ϕ is less than the *critical angle* θ_c, which can be derived from Snell's law

$$n_1 \sin\theta_i = n_2 \sin\theta_c; \qquad \theta_c = \arcsin\frac{n_2}{n_1}. \tag{1.76}$$

If this condition is not satisfied, the ray is no longer confined in the waveguide and radiates into the cladding, where it is lost. Note that when the waveguide is curved or has a rough interface between core and cladding, there is a greater chance that the ray will meet the interface at an obtuse angle and be lost. Hence in order to achieve low-loss waveguides, we should engineer smooth interfaces and gentle curves. From this intuitive picture we can also see that a greater *refractive-index contrast*

$$\Delta n = \frac{n_1^2 - n_2^2}{2n_1^2} \tag{1.77}$$

between core and cladding leads to a larger critical angle, allowing tighter bends and thus smaller, more compact structures.

Given a specific device geometry, the Helmholtz Eq. 1.68 is only satisfied only for a discrete subset of spatial distributions $\mathbf{E}_0$, referred to as *waveguide modes*. Owing to the complexity and breadth of possible device geometries, the form of these modes must in general be calculated using numerical mode solvers (FIMMWave [41], Phoenix [42] etc.), but for a simple one-dimensional model we can find an analytic solution.

Consider for example the refractive index profile shown in Fig. 1.4 for a waveguide of width $2a$

$$n = n_1 \qquad |x| < a \tag{1.78}$$

$$n = n_2 \qquad |x| \geq a. \tag{1.79}$$

Taking electric field propagation to be in the *transverse electric* (TE) mode, which for a particular choice of coordinate system is equivalent to saying that E_y is the only nonzero component of $\mathbf{E}_0$, (1.68) becomes

$$\frac{\partial E_y}{\partial x^2} = \gamma^2 E_y \quad |x| < a; \qquad \frac{\partial E_y}{\partial x^2} = -\kappa^2 E_y \quad |x| \geq a \tag{1.80}$$

in the core and cladding respectively, where γ^2 and κ^2 are real parameters which depend on both on the structure and material of the waveguides, and on the wavelength of incident light. These equations have solutions of the form

$$E_y = G_1 e^{\gamma x} \qquad x \leq -a \tag{1.81}$$

$$E_y = G_2 e^{i\kappa x} G_3 e^{-i\kappa x} \qquad -a < x < a \tag{1.82}$$

$$E_y = G_4 e^{-\gamma x} \qquad x \geq a \tag{1.83}$$

where G_i are constants which depend on the waveguide parameters and the optical wavelength. This captures an important property of optical waveguides which is not described by the ray-optics model: Fig. 1.4b suggests that under total internal reflection the optical field is always fully confined within the core and does not impinge on the cladding, whereas in practise this is not the case. We see from (1.81) and (1.83) that outside the waveguide core the electromagnetic field is not zero, instead falling off exponentially with distance. This is the *evanescent field* of the waveguide mode, which permits two waveguides to be coupled together without bringing the cores into contact. This is discussed in further detail in Sect. 2.2.2.

We have already seen (1.73) that the contrast of optical interference is reduced when the two sources have differing polarization. By a similar argument, in order to see high-contrast interference between light sources in guided modes, we should engineer the waveguide, through control of the geometry, size, and refractive index, so as to support only a single guided mode—a single solution of (1.68)—for a target wavelength λ. These are known as *single-mode* (as opposed to *multimode*) waveguides, and are used throughout this thesis.

1.5.2 Light as a Photon

Before examining the quantum-mechanical description of light, it will helpful to revise the properties of the classical and quantum harmonic oscillators. In a classical simple harmonic oscillator (SHO), such as a spring or pendulum with spring constant k, the force acting on a particle is proportional to its displacement, $F = -kx$. The dynamics are described by the classical SHO Hamiltonian

$$KE = \int F \cdot v\,dt = \frac{1}{2}mv^2 = \frac{p^2}{2m}; \qquad PE = \int k \cdot x\,dx = \frac{1}{2}kx^2 = \frac{m\omega^2}{2} \tag{1.84}$$

$$H = KE + PE = \frac{p^2}{2m} + \frac{m\omega^2 x^2}{2} \tag{1.85}$$

where $\omega = \sqrt{k/m} = 2\pi f$ is the angular frequency. From the Hamilton equations $\dot{p} = -\frac{\partial H}{\partial x}$, $\dot{x} = +\frac{\partial H}{\partial p}$, we arrive at the SHO equation of motion

$$\frac{d^2x}{dt^2} = -\omega^2 x \tag{1.86}$$

In close analogy with the general solution of Maxwell's equations (1.67), a general solution to (1.86) is $\alpha(t) = \alpha(0)e^{-i\omega t}$, an unphysical (complex) ansatz. Just as **E**, **B** are related to the real and imaginary parts of $\alpha_{\mathbf{k}}$ for the electromagnetic field, the complex components of $\alpha(t)$ are mapped by convention to the position and momentum of the SHO, respectively:

$$x(t) = \sqrt{\frac{\hbar}{2m\omega}}\left[\alpha + \alpha^*\right] \propto Re\left[\alpha(t)\right]; \quad p(t) = i\sqrt{\frac{\hbar m\omega}{2}}\left[\alpha^* - \alpha\right] \propto Im\left[\alpha(t)\right] \tag{1.87}$$

$$\alpha(t) = \frac{1}{\sqrt{2\hbar}}\left[\sqrt{m\omega}x(t) + \frac{i}{\sqrt{m\omega}}p(t)\right]. \tag{1.88}$$

Hence α can be thought of as providing a compact phase-space representation of the state of the SHO, (x, p). Here we have assumed that $\alpha(t)$ is dimensionless, allowing us to rescale by $\hbar$, a constant with units of action. The SHO Hamiltonian (1.85) can then be re-written in terms of $\alpha(t)$ as

$$H = \frac{\hbar\omega}{2}\left[\alpha^*(t)\alpha(t) + \alpha(t)\alpha^*(t)\right]. \tag{1.89}$$

We now turn to the quantum harmonic oscillator (QHO). The state of a quantum particle is represented by a state vector $|\psi(x)\rangle$ in Hilbert space, and the position and momentum observables become non-commuting Hermitian operators acting on this space

$$\hat{x} = x; \qquad \hat{p} = -i\hbar\frac{\partial}{\partial x} \tag{1.90}$$

with $\left[\hat{x}, \hat{p}\right] = i\hbar$ (see Sect. 1.3.4). As with the SHO (1.85), the QHO Hamiltonian is then given by

$$\hat{H} = \frac{\hat{p}^2}{2m} + \frac{m\omega^2\hat{x}^2}{2} \tag{1.91}$$

and $|\psi\rangle$ satisfies the time-independent Schrödinger equation, $\hat{H}|\psi\rangle = E|\psi\rangle$. The QHO has an analogous solution to that of the SHO (1.85), where $\alpha, \alpha^\dagger$ are replaced by their quantized counterparts

$$\hat{a} = \sqrt{\frac{m\omega}{2\hbar}}\left(\hat{x} + \frac{i}{m\omega}\hat{p}\right); \quad \hat{a}^\dagger = \sqrt{\frac{m\omega}{2\hbar}}\left(\hat{x} - \frac{i}{m\omega}\hat{p}\right) \tag{1.92}$$

leading to

$$\hat{x} = \sqrt{\frac{\hbar}{2m\omega}}\left(\hat{a} + \hat{a}^\dagger\right); \quad \hat{p} = i\sqrt{\frac{\hbar m\omega}{2}}\left(\hat{a}^\dagger - \hat{a}\right). \tag{1.93}$$

The operators $\hat{a}^\dagger$ and $\hat{a}$ are known as the *creation* and *annihilation* operators for the QHO, respectively. $\hat{a}, \hat{a}^\dagger$ are not real observables, and are therefore not Hermitian. In contrast with the classical case, however, they do not commute, with $\left[\hat{a}, \hat{a}^\dagger\right] = 1$. They are jointly named the *ladder operators*, since their action on an energy eigenstate is to raise or lower the energy by a single quantum $\hbar\omega$,

$$\hat{a}^\dagger|n\rangle = \sqrt{n+1}|n+1\rangle; \quad \hat{a}|n\rangle = \sqrt{n}|n-1\rangle; \quad \hat{a}|0\rangle = 0. \tag{1.94}$$

We can also define the *number operator* $\hat{N} \equiv \hat{a}^\dagger\hat{a}$, which "counts" the number of quanta in an energy eigenstate, $\hat{N}|n\rangle = n|n\rangle$.

Using (1.93) together with the commutation relation $\hat{a}\hat{a}^\dagger = \hat{a}^\dagger\hat{a} + 1$, the QHO Hamiltonian (1.91) can be re-written as

$$\hat{H} = \frac{\hbar\omega}{2}\left(\hat{a}^\dagger\hat{a} + \hat{a}\hat{a}^\dagger\right) = \hbar\omega\left(\hat{a}^\dagger\hat{a} + \frac{1}{2}\right), \tag{1.95}$$

which has eigenstates $|n\rangle$ of energy $E_n = \hbar\omega\left(n + \frac{1}{2}\right), n \in \{\mathbb{Z} : n \geq 0\}$. Note that the energy of the QHO ground state $|0\rangle$ is not zero, $E_0 = \frac{1}{2}\hbar\omega > 0$.

We now proceed to quantization of the electromagnetic field. We first note the similarity between the Hamiltonian of the linear electromagnetic field in a mode $\mathbf{k}$ (1.71) in terms of an ansatz $\alpha_{\mathbf{k}}(t)$, and the QHO Hamiltonian (1.95) in terms of the annihilation operator $\hat{a}$. Similarly, there is a corresponence between the position and momentum operators $\hat{x}, \hat{p}$ and the electric and magnetic fields, $\mathbf{E}, \mathbf{B}$. This allows us to take an analogous approach to the SHO, replacing $\alpha_{\mathbf{k}}$ and $\alpha^*_{\mathbf{k}}$ by the ladder operators $\hat{a}^\dagger_{\mathbf{k}}, \hat{a}_{\mathbf{k}}$ acting on a mode $\mathbf{k}$, and choosing a dispersion relation $\omega = ck$:

$$H^{EMF}_{\mathbf{k}} = \frac{\hbar ck}{2}\left(\alpha^*_{\mathbf{k}}\alpha_{\mathbf{k}} + \alpha_{\mathbf{k}}\alpha^*_{\mathbf{k}}\right); \quad \hat{H}^{QHO} = \frac{\hbar\omega}{2}\left(\hat{a}^\dagger\hat{a} + \frac{1}{2}\right) \tag{1.96}$$

$$\rightarrow \quad \hat{H}_{\mathbf{k}} = \hbar\omega\left(\hat{a}^\dagger_{\mathbf{k}}\hat{a}_{\mathbf{k}} + \frac{1}{2}\right) \tag{1.97}$$

Now, the state of the electromagnetic field is represented by a vector $|\psi\rangle$ in Hilbert space $\mathscr{H}$, and $\hat{a}^\dagger_{\mathbf{k}}$, $\hat{a}_{\mathbf{k}}$ are ladder operators acting on $\mathscr{H}$ which create or destroy a *photon* of energy $\hbar\omega$, respectively:

$$\hat{a}^\dagger_{\mathbf{k}}|n\rangle_{\mathbf{k}} = \sqrt{n+1}|n+1\rangle_{\mathbf{k}}; \qquad \hat{a}_{\mathbf{k}}|n\rangle_{\mathbf{k}} = \sqrt{n}|n-1\rangle_{\mathbf{k}}; \qquad \hat{a}_{\mathbf{k}}|0\rangle_{\mathbf{k}} = 0. \quad (1.98)$$

The eigenstates $|n\rangle_{\mathbf{k}}$ of the quantized electromagnetic Hamiltonian (1.97) are called the *number* or *Fock* states,[4] and form an orthonormal basis for $\mathscr{H}$. A mode $\mathbf{k}$ in Fock state $|n\rangle_{\mathbf{k}}$ is interpreted as literally containing $\langle n|\hat{a}^\dagger_{\mathbf{k}}\hat{a}_{\mathbf{k}}|n\rangle = n$ photons, $n \in \mathbb{Z}$. Note that a mode containing zero photons still has nonzero energy, $E_0 = \hbar\omega/2$: this is the *vacuum energy* of the electromagnetic field. Any Fock state can be written in terms of the vacuum state $|0\rangle_{\mathbf{k}}$,

$$|n\rangle_{\mathbf{k}} = \frac{1}{\sqrt{n!}}(\hat{a}^\dagger_{\mathbf{k}})^n|0\rangle_{\mathbf{k}}. \quad (1.99)$$

and a general superposition state in mode $\mathbf{k}$ can be written in the Fock basis

$$|\psi\rangle_{\mathbf{k}} = \sum_{n=0}^{N} b_n|n\rangle_{\mathbf{k}}. \quad (1.100)$$

To summarize, we have seen that quantization of the electromagnetic field in a single mode $\mathbf{k}$ leads to solutions which are strongly analogous to the energy eigenstates of the quantum harmonic oscillator, corresponding to the Fock states $|n\rangle$ of n photons, each with energy $\hbar\omega$. All of the experiments described in this thesis, together with most quantum photonic technologies, depend on the use of many photons in many modes, In the next section we outline basic notation and methods used to deal with such states, as well as some associated physical phenomena.

1.5.2.1 Photons in Modes

Our discussion so far has been limited to the creation and annihilation of photons in a single spatial mode $\mathbf{k}$. The experimental work presented in this thesis, however, deals with $2 \le p \le 6$ photons in $2 \le m \le 21$ modes, and makes use of both time and polarization degrees of freedom. In order to provide a more complete framework, we map $(\hat{a}^\dagger_{\mathbf{k}}, \hat{a}_{\mathbf{k}}) \rightarrow (\hat{a}^\dagger_j, \hat{a}_j)$ where j indexes any allowed field mode of the system, including modes in time, space, frequency and polarization. We will principally be concerned with photons which are *indistinguishable* in the sense that any two photons can be swapped in any experimental degree of freedom without changing the state

[4]After V.A. Fock, whose name is also given to the *Hartree-Fock* method described in Sect. 5.3.2.

of the overall system. Indistinguishable bosons in modes i, j obey the *canonical commutation relations*

$$\left[\hat{a}_i, \hat{a}_j^\dagger\right] = \delta_{ij}\mathbf{1}; \qquad \left[\hat{a}_i, \hat{a}_j\right] = \left[\hat{a}_i^\dagger, \hat{a}_j^\dagger\right] = 0, \tag{1.101}$$

which capture many important properties of the photonic ladder operators, and will be useful throughout this discussion.

The Hilbert space $\mathscr{H}_m^p$ for the state of p indistinguishable photons in m modes is generated by the tensor product (see Sect. 1.3.1), and we write the eigenstates of an arbitrary number of photons $p = \sum_j n_j$ occupying m modes in Fock notation as

$$|n\rangle_1 \otimes |n\rangle_2 \ldots |n\rangle_m = |n_1, n_2, \ldots n_m\rangle = \left[\prod_{j=1}^{m} \frac{1}{\sqrt{n_j!}} \left(\hat{a}_j^\dagger\right)^{n_j}\right] |\mathbf{0}\rangle \tag{1.102}$$

where $|\mathbf{0}\rangle \equiv |0\rangle_0 \otimes |0\rangle_1 \ldots |0\rangle_n = |00\ldots0\rangle$ is the m-mode vacuum. These states form an orthonormal basis for the Hilbert space $\mathscr{H}_m^p$ of p photons in m modes, and an arbitrary pure superposition state can therefore be written as

$$|\psi\rangle = \sum_i^d b_i |n_{1,i}, n_{2,i} \ldots n_{m,i}\rangle = \left[\sum_i^d b_i \prod_{j=1}^{m} \frac{1}{\sqrt{n_{ij}!}} \left(\hat{a}_j^\dagger\right)^{n_{ij}}\right] |\mathbf{0}\rangle\,, \tag{1.103}$$

where d is the Hilbert space dimension and $\sum_i |b_i|^2 = 1$. Many experiments in quantum optics deal with a large number of modes and a *fixed* number of photons. The Hilbert space dimension d of $\mathscr{H}_m^p$ corresponds to the number of unique configurations of p indistinguishable photons in m modes, given by the binomial coefficient

$$d = \binom{m+p-1}{p}; \qquad D = \binom{m}{p}, \tag{1.104}$$

where $D < d$ is the dimension of the *collision-free subspace* in which no two photons occupy the same mode.

1.5.2.2 The Coherent State

The *coherent state* $|\alpha\rangle$ is the state of the quantized electromagnetic field whose dynamics most resemble a classical harmonic oscillator. It provides a good approximation to the state generated by a continuous-wave laser—an essentially classical state of light. It will be important to contrast the behaviour of the coherent state against that of Fock states, in order to motivate the use of single-photon sources throughout this thesis. Here we follow Glauber [43].

The coherent state is defined as an eigenstate of the annihilation operator $\hat{a}$ with eigenvalue α,

$$\hat{a}|\alpha\rangle = \alpha|\alpha\rangle. \tag{1.105}$$

Expressing $|\alpha\rangle$ in the Fock basis (1.100), $|\alpha\rangle = \sum_{n=0}^{\infty} b_n |n\rangle$, we can re-write (1.105)

$$\sum_{n=1}^{\infty} b_n \sqrt{n} |n-1\rangle = \alpha \sum_{n=0}^{\infty} b_n |n\rangle \tag{1.106}$$

and by re-indexing the left hand side,

$$\sum_{n=0}^{\infty} b_{n+1} \sqrt{n+1} |n\rangle = \alpha \sum_{n=0}^{\infty} b_n |n\rangle \quad \rightarrow \quad b_{n+1} = \frac{\alpha}{\sqrt{n+1}} b_n \quad \text{for } n \geq 0. \tag{1.107}$$

This recursive expression provides the superposition coefficients $b_n = \frac{\alpha^n}{\sqrt{n!}} b_0$. Since the state must be normalized $\langle\alpha|\alpha\rangle = 1$, we have $1/|b_0|^2 = e^{|\alpha|^2}$. Choosing the phase of b_0 so as to make it real, we arrive at a Fock-basis form for the coherent state,

$$|\alpha\rangle = e^{-\frac{|\alpha|^2}{2}} \sum_{n=0}^{\infty} \frac{\alpha^n}{\sqrt{n!}} |n\rangle = e^{\alpha \hat{a}^\dagger - \alpha^* \hat{a}} |\mathbf{0}\rangle. \tag{1.108}$$

The coherent state $|\alpha\rangle$ has average photon number $\langle n\rangle = |\alpha|^2$, but in contrast with the Fock state $|n\rangle$ there is a nonzero probability of detecting more than n photons simultaneously. In general, the probability $P(n)$ of detecting photon number n from $|\alpha\rangle$ has a Poissonian distribution:

$$P(n) = |\langle n|\alpha\rangle|^2 = e^{-|\alpha|^2} \frac{|\alpha|^{2n}}{n!}. \tag{1.109}$$

The normalized second-order correlation function for photons generated in a single spatial mode at times $t = 0, t + \tau$

$$g^{(2)}(\tau) = \frac{\langle \hat{a}_0^\dagger \hat{a}_\tau^\dagger \hat{a}_\tau \hat{a}_0\rangle}{\langle \hat{a}_0^\dagger \hat{a}_0\rangle \langle \hat{a}_\tau^\dagger \hat{a}_\tau\rangle}; \quad g^{(2)}(0) = \frac{\langle (\hat{a}^\dagger)^2 \hat{a}^2\rangle}{\langle \hat{a}^\dagger \hat{a}\rangle^2} = \frac{\mathrm{var}(n) - \langle n\rangle}{\langle n\rangle^2} + 1, \tag{1.110}$$

characterises the relationship between the mean and variance of the photon number distribution, and is similar but not equivalent to the classical cross-correlation function (6.2). For the coherent state,

$$g^{(2)}(0) = \frac{\langle (\hat{a}^\dagger)^2 \hat{a}^2\rangle}{\langle \hat{a}^\dagger \hat{a}\rangle^2} = \frac{\langle \alpha|\alpha^* \hat{N} \alpha|\alpha\rangle}{\langle \alpha|\alpha^* \alpha|\alpha\rangle} = \frac{|\alpha|^4}{|\alpha|^4} = 1 \tag{1.111}$$

However, for a Fock state $|n\rangle$, $g^{(2)}(0)$ is less than unity

$$g^{(2)}(0) = \frac{\langle(\hat{a}^\dagger)^2\hat{a}^2\rangle}{\langle\hat{a}^\dagger\hat{a}\rangle^2} = \frac{\langle n|(\hat{N}-1)\hat{N}|n\rangle}{\langle n|\hat{N}|n\rangle^2} = 1 - \frac{1}{n} \leq 1. \tag{1.112}$$

For a given light source, if $g^{(2)}(0) < 1$ then the photon-number distribution $P(n)$ has a smaller variance than the equivalent Poisson distribution, and the source is said to be *sub-Poissonian* and *nonclassical*. This effect is referred to as photon *antibunching*, in the sense that it is unlikely or impossible for many photons to arrive simultaneously at the detector. For incoherent (chaotic) light the opposite is true, and the twofold detection probability is instead enhanced with respect to that of statistically independent particles, giving $g^{(2)} > 1$. This is the *Hanbury-Brown-Twiss* [44] effect, and is referred to as *bunching*, since photons appear to clump together upon arrival.

1.5.2.3 Time Evolution of Photons

General methods for time evolution of quantum states are discussed in Sect. 1.3.3. In this thesis, time evolution is almost always due to a linear-optical *circuit*—a static, discrete network of waveguides and/or bulk optical elements, which takes an input state $|\psi\rangle_{\text{in}}$ to an output state $|\psi\rangle_{\text{out}}$. A lossless, time-independent circuit can always be completely described by unitary matrix $\hat{U}$ which maps between the input and output modes of the device, labelled a_i and b_j respectively.

Starting from a general pure input state in the form of (1.103), we can study time-evolution in the Heisenberg picture, writing

$$|\psi\rangle_{\text{out}} = \hat{U}|\psi\rangle_{\text{in}} = \hat{U}\left[\sum_i b_i \prod_{j=1}^{m} \frac{\left(\hat{a}^\dagger_{a_j}\right)^{n_{ij}}}{\sqrt{n_{ij}!}}\right]\hat{U}^\dagger\hat{U}|\mathbf{0}\rangle = \left[\sum_i b_i \prod_{j=1}^{m} \frac{\left(\hat{U}\hat{a}^\dagger_{a_j}\hat{U}^\dagger\right)^{n_{ij}}}{\sqrt{n_{ij}!}}\right]|\mathbf{0}\rangle \tag{1.113}$$

where we have used the fact that $\hat{U}|\mathbf{0}\rangle = e^{i\phi}|\mathbf{0}\rangle \rightarrow |\mathbf{0}\rangle$ (optical circuits described by unitary operators do not create or destroy photons, and the global phase is unobservable) and $\hat{U}\hat{U}^\dagger = \hat{U}^\dagger\hat{U} = \mathbf{1}$. Now, the output-mode creation operators can be written in terms of the input fields

$$\hat{a}^\dagger_{b_j} = \hat{U}\hat{a}^\dagger_{b_j}\hat{U}^\dagger. \tag{1.114}$$

The time-evolution of general multiphoton states can thus be computed based on a model of the *single-particle* statistics, $\hat{U}$. Since single-photon solutions of the Heisenberg equation have identical solutions to the classical field, this allows us to model general multiphoton behaviour starting from a classical understanding of the system. The unitary $\hat{U}$, which completely and uniquely characterizes the circuit, can

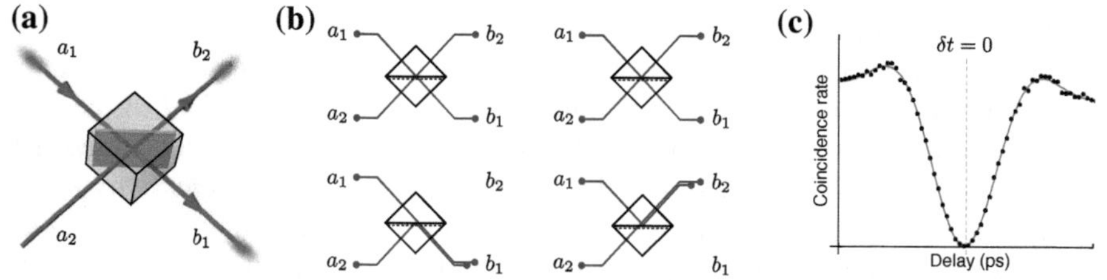

Fig. 1.5 **a** A bulk-optical beamsplitter is modelled as having two single-mode input ports a_1, a_2 and two output ports b_1, b_2. If bright light is injected into input port a_1, a fraction r of the total light intensity will be reflected into output port b_2, while $t = 1 - r$ is transmitted to b_1. **b** In a Hong-Ou-Mandel interference experiment, two indistinguishable photons are sent into ports a_1, a_2 of a BS. There are four possible outcomes of the experiment: both photons can be transmitted, both reflected, one transmitted and one reflected, and vice-versa. **c** When the photons are perfectly indistinguishable, the first two measurement outcomes destructively interfere and the probability of coincidental detection at b_1, b_2 vanishes. By tuning the distinguishability of the photons, we can map out the *Hong-Ou-Mandel dip* in the coincidence rate. Experimental data is used here only for illustration purposes, and is shown complete with error bars and accidental coincidence count-rates in Fig. 2.10

always be represented as an $m \times m$ matrix, where in general m is much smaller than the Hilbert space dimension of $\mathscr{H}_m^p$.

Note that although these calculations can be performed based on a classical starting-point, that is not to say that all of the resulting multiphoton behaviour can be explained by a classical model, as we will see in the next section. Furthermore, there is strong evidence to suggest that not all states and probabilities generated by (1.113) can be efficiently calculated on a classical computer—in many cases the number of terms in the expansion is exponentially large in p. See Sects. 1.5.3.2 and 6.3.2 for further discussion of this point.

It will often be convenient to re-write (1.114) for the input field operators in terms of the output fields and a unitary matrix $\mathbf{\Lambda}$, which is analogous to the classical transfer matrix

$$\hat{a}_{a_i}^\dagger = \sum_j^m \mathbf{\Lambda}_{ij}\, \hat{a}_{b_j}^\dagger; \quad \mathbf{\Lambda}^\dagger \mathbf{\Lambda} = \mathbf{1}. \tag{1.115}$$

1.5.2.4 The Beamsplitter

The beamsplitter (BS), shown schematically in Fig. 1.5a, is a basic component of optical circuits. The most common design of a bulk-optical BS consists of two triangular prisms of BK-7 borosilicate glass, glued together with the resin of a fir tree[5] so as to form a cube with a plane interface across the main diagonal. Half-silvered

[5] The Canada balsam fir, *Abies balsamea*.

mirrors, microscope slides, and integrated optics (Sect. 1.6), amongst others, can all be used to construct effective beamsplitters. A light beam incident at 45° to the interface is split into two orthogonal output modes, with a fraction $r = I_r/I$ of the input intensity reflected at 90° to the incident beam, and $t = I_t/I = 1-r$ transmitted. The BS is thus completely characterized by the *reflectivity* r and *transmissivity* t, also referred to as the *coupling ratio* $\eta = t$. A *50:50 BS* is designed to have $r = t = \frac{1}{2}$.

If a classical light field is injected into one of the two input modes a_1, a_2, the effect of the BS is to split the complex amplitude α of the input field across the two output modes, conserving energy and momentum, as

$$\alpha_{b_1} = \alpha_{a_1}\sqrt{t} + i\alpha_{a_2}\sqrt{r}, \qquad \alpha_{b_2} = i\alpha_{a_1}\sqrt{r} + \alpha_{a_2}\sqrt{t}, \tag{1.116}$$

$$\rightarrow \quad \alpha_{a_1} = \alpha_{b_1}\sqrt{t} - i\alpha_{b_2}\sqrt{r}, \qquad \alpha_{a_2} = -i\alpha_{b_1}\sqrt{r} + \alpha_{b_2}\sqrt{t}. \tag{1.117}$$

Here the factor i arises on reflection, and is necessary for energy to be conserved. The details of this kōan of experimental optics, “the photon picks up a phase on reflection”, are not often discussed, and the effect is less obvious than it might seem. Certainly, we could build an optical element, resembling a beamsplitter, whose effect is characterised exactly by a Hadamard matrix—in which case it is not always the case that the photon picks up a phase on reflection. Full analysis, given for example in [45], is outside the scope of this thesis. The relations (1.117) lead directly to the quantum beamsplitter transformation for ladder operators in the Heisenberg picture

$$\hat{a}^\dagger_{a_1} \xrightarrow{BS} \hat{a}^\dagger_{b_1}\sqrt{t} + i\hat{a}^\dagger_{b_2}\sqrt{r}, \quad \hat{a}^\dagger_{a_2} \xrightarrow{BS} i\hat{a}^\dagger_{b_1}\sqrt{r} + \hat{a}^\dagger_{b_2}\sqrt{t}. \tag{1.118}$$

The beamsplitter has an associated unitary operator $\hat{U}_{\mathrm{BS}}$ as well as a $\mathbf{\Lambda}$-matrix,

$$\mathbf{\Lambda}_{BS}(r) = \begin{bmatrix} \sqrt{t} & i\sqrt{r} \\ i\sqrt{r} & \sqrt{t} \end{bmatrix} \tag{1.119}$$

allowing (1.118) to be re-written as

$$\begin{bmatrix} \hat{a}^\dagger_{a_1} \\ \hat{a}^\dagger_{a_2} \end{bmatrix} = \begin{bmatrix} \sqrt{t} & i\sqrt{r} \\ i\sqrt{r} & \sqrt{t} \end{bmatrix} \begin{bmatrix} \hat{a}^\dagger_{b_1} \\ \hat{a}^\dagger_{b_2} \end{bmatrix} = \begin{bmatrix} \hat{a}^\dagger_{b_1}\sqrt{t} + i\hat{a}^\dagger_{b_2}\sqrt{r} \\ i\hat{a}^\dagger_{b_1}\sqrt{r} + \hat{a}^\dagger_{b_2}\sqrt{t} \end{bmatrix}. \tag{1.120}$$

Let’s compare the behaviour of single photons incident on a 50:50 BS with that of the coherent state. If we inject a single photon into mode a_1, the system evolves as

$$\hat{a}^\dagger_{a_1}|\mathbf{0}\rangle \xrightarrow{BS} \hat{U}_{\mathrm{BS}}\hat{a}^\dagger_{a_1}\hat{U}^\dagger_{\mathrm{BS}}|\mathbf{0}\rangle = \frac{1}{\sqrt{2}}(\hat{a}^\dagger_{b_1} + i\hat{a}^\dagger_{b_2})|\mathbf{0}\rangle = \frac{1}{\sqrt{2}}\left(|1_{b_1}0_{b_2}\rangle + i|0_{b_1}1_{b_2}\rangle\right). \tag{1.121}$$

Note that the photon is only ever detected in one or other of the output ports, never both at the same time ($\langle\psi|11\rangle = 0$). This is the effect of photon antibunching (1.112) which was first experimentally confirmed in 1986 by Grangier et al. [46], constituting arguably the first strong evidence for fully particle-like behaviour of the photon. It is interesting to note that when written in the Fock basis, (1.121) is locally equivalent to a Bell state (1.38). See Ref. [47] for further discussion of entanglement and nonlocality of a single photon.

If we instead inject a single coherent state $|\alpha\rangle_{a_1}$ at input port a_1, the output state is

$$\hat{U}_{\mathrm{BS}}|\alpha\rangle_{a_1} = \exp\left[\frac{(\alpha\hat{a}^\dagger_{b_1} - \alpha^*\hat{a}_{b_1}) + i(\alpha\hat{a}^\dagger_{b_2} + \alpha^*\hat{a}_{b_2})}{\sqrt{2}}\right]|\mathbf{0}\rangle = \frac{i}{\sqrt{2}}|\alpha\rangle_{b_1} \otimes |\alpha\rangle_{b_2} \tag{1.122}$$

which *does* have nonzero $|11\rangle$ terms ($\langle\psi|11\rangle \neq 0$), allowing two photons to be coincidentally detected at both output ports and leading to $g^{(2)}(0) > 1$ (1.111).

Although the single photon and the coherent state are distinguished by correlated detection statistics (i.e. antibunching), in non-correlated measurements they give essentially identical measurement outcomes. For example, the probability that a single detector will fire at either output port of a beamsplitter is the same for both a single- photon source and a coherent state $|\alpha = 1\rangle$,

$$P(n_{b_1} \geq 1) = P(n_{b_2} \geq 1) = \frac{1}{2}. \tag{1.123}$$

1.5.3 Quantum Interference

To see a stronger distinction between quantum and classical behaviour of photons, we now consider a situation in which multiple light sources are used, rather than one. Dirac famously addressed experiments of this type, arguing that since interference between different sources would seem to involve the creation or destruction of photons, violating conservation of energy, it should not occur:

> Each photon then interferes only with itself. Interference between two different photons never occurs.
>
> P.A.M. Dirac, *The Principles of Quantum Mechanics* [3]

We will now show that this intuition, which is supported by our everyday experience of the behaviour of light, does not always hold. Specifically, we will see that two indistinguishable photons launched into different ports of a 50:50 BS interfere with one another, precluding simultaneous detection of photons at two output ports—an effect which has no classical analogue.

Quantum interference, as this effect is known, is thus the basic mechanism that we will use to allow one photon to "talk" to another. It is used throughout this thesis to implement entangling gate operations on path-encoded photonic qubits, and is essential for linear-optical quantum computing (discussed in Sect. 1.6.2) as well as the "boson computer" (Sect. 6.3.2). In Sect. 6.3, we observe generalized quantum interference between up to 5 photons in 21 spatial modes.

1.5.3.1 Two-Photon Interference

Consider the situation shown in Fig. 1.5b, in which two single photons are sent into the input ports of a 50:50 BS (a_1, a_2 respectively). We assume that the photons are indistinguishable in all degrees of freedom apart from path, having the same polarization, wavelength etc., For classical particles, this experiment has four possible outcomes: both particles can be transmitted, both reflected, one transmitted and one reflected, and vice versa. Since there are no interference effects for classical particles, the detection probability at output ports (i, j) is simply given by the product of the corresponding single-particle probabilities, $P(i \cap j) = P(ij) = P(i) \cdot P(j)$,

$$P(2_{b_1}0_{b_2}) = P(b_1b_1) = P(b_1) \cdot P(b_1) = \frac{1}{2} \cdot \frac{1}{2} = \frac{1}{4}; \quad P(0_{b_1}2_{b_2}) = P(b_2b_2) = \frac{1}{4} \tag{1.124}$$

$$P(1_{b_1}1_{b_2}) = P(b_1b_2 \cup b_2b_1) = P(b_1b_2) + P(b_2b_1) = \frac{1}{4} + \frac{1}{4} = \frac{1}{2}. \tag{1.125}$$

For photons, the output state of the BS is given by

$$|1_{a_1}1_{a_2}\rangle = \hat{a}^\dagger_{a_1}\hat{a}^\dagger_{a_2}|\mathbf{0}\rangle \quad \xrightarrow{BS} \quad \frac{1}{2}\left(\hat{a}^\dagger_{b_1} + i\hat{a}^\dagger_{b_2}\right)\left(i\hat{a}^\dagger_{b_1} + \hat{a}^\dagger_{b_2}\right)|\mathbf{0}\rangle \tag{1.126}$$

$$= \frac{i}{2}\left((\hat{a}^\dagger_{b_1})^2 - \hat{a}^\dagger_{b_1}\hat{a}^\dagger_{b_2} + \hat{a}^\dagger_{b_2}\hat{a}^\dagger_{b_1} + (\hat{a}^\dagger_{b_2})^2\right)|\mathbf{0}\rangle. \tag{1.127}$$

Using the canonical commutation relations (1.101) this becomes

$$|\psi\rangle_{\text{out}} = \frac{1}{2}\left((\hat{a}^\dagger_{b_1})^2 + (\hat{a}^\dagger_{b_2})^2\right)|\mathbf{0}\rangle = \frac{1}{\sqrt{2}}\left(|2_{b_1}0_{b_2}\rangle + |0_{b_1}2_{b_2}\rangle\right), \tag{1.128}$$

where we have ignored the global phase i (which cannot be measured). We then have

$$P(2_{b_1}0_{b_2}) = |\langle 2_{b_1}0_{b_2}|\psi\rangle|^2 = \frac{1}{2}; \quad P(0_{b_1}2_{b_2}) = \frac{1}{2}; \quad P(1_{b_1}1_{b_2}) = 0. \tag{1.129}$$

Thus the probability that two photons are simultaneously detected at different output ports, the probability of coincidental detection, vanishes, This is in strong contrast with the behaviour of classical particles (1.125), and can only be explained by interference between the two sources. This is the famous Hong-Ou-Mandel (HOM) interference effect, also known as two-photon quantum interference, first proposed and experimentally demonstrated in 1987 by Hong et al. [48].

Note that the state (1.128) is not separable—it cannot be written as a product state of the two systems as (1.35), and is therefore entangled. This a *NOON* state, which can be used to achieve quantum-enhanced precision in measurements, as discussed in Sect. 1.4.3.

For distinguishable photons, the situation is comparable to that of classical particles. To see this, let each mode a_i now be associated with *two* modes (a_i, a_i'), which are distinguishable (orthogonal) in, for instance, polarization or time. Now, the system evolves as

$$|1_{a_1}0_{a_1'}0_{a_2}1_{a_2'}\rangle = \hat{a}^\dagger_{a_1}\hat{a}^\dagger_{a_2'}|\mathbf{0}\rangle \quad \stackrel{BS}{\longrightarrow} \quad \frac{i}{2}\left(\hat{a}^\dagger_{b_1}\hat{a}^\dagger_{b_1'} + \hat{a}^\dagger_{b_1}\hat{a}^\dagger_{b_2'} - \hat{a}^\dagger_{b_2}\hat{a}^\dagger_{b_1'} + \hat{a}^\dagger_{b_2}\hat{a}^\dagger_{b_2'}\right)|\mathbf{0}\rangle. \tag{1.130}$$

Since the two photons can in principle be distinguished by this extra degree of freedom, the creation operators no longer commute $\hat{a}^\dagger_{b_1}\hat{a}^\dagger_{b_2'} \neq \hat{a}^\dagger_{b_2}\hat{a}^\dagger_{b_1'}$ and the output state is then

$$|\psi\rangle_{\text{out}} = \frac{i}{2}\Big(|1_{b_1}1_{b_1'}0_{b_2}0_{b_2'}\rangle + |1_{b_1}0_{b_1'}0_{b_2}1_{b_2'}\rangle - |0_{b_1}1_{b_1'}1_{b_2}0_{b_2'}\rangle + |0_{b_1}0_{b_1'}1_{b_2}1_{b_2'}\rangle\Big). \tag{1.131}$$

Then, tracing over the orthogonal modes, we recover classical particle statistics (1.125), with nonzero probability of coincidental detection at separate output ports:

$$P(2_{b_1}0_{b_2}) = |\langle 1_{b_1}1_{b_2'}|\psi\rangle|^2 = \frac{1}{4}; \quad P(0_{b_1}2_{b_2}) = \frac{1}{4}; \tag{1.132}$$

$$P(1_{b_1}1_{b_2}) = |\langle 1_{b_1}1_{b_2'}|\psi\rangle|^2 + |\langle 1_{b_1'}1_{b_2}|\psi\rangle|^2 = \frac{1}{2}. \tag{1.133}$$

In experimental demonstrations of quantum interference, the average coincidence count-rate $c(1_{b_1}1_{b_2}) = C \cdot P(1_{b_1}1_{b_2})$ is very often measured as a continuous function of the distinguishability of the photon pair, where C is the total count-rate across all detection patterns. By controlling the arrival time (as in [48] and in this thesis) or polarization of one photon with respect to the other, a so-called *HOM dip* in coincidences can be mapped out. Figure 1.5c shows a HOM dip measured using an integrated beamsplitter (Sect. 2.2.2), in which the pair distinguishability is tuned by delaying the arrival time of one photon with respect to the other, on the order of the coherence time of the photon (picoseconds). When the delay is much greater than the coherence time, the photons are fully distinguishable and $P(1_{b_1}1_{b_2}) = 1/2$, while for zero delay, the photons are maximally indistinguishable and $P(1_{b_1}1_{b_2}) \to 0$. The shape of the dip depends on various properties of the photons themselves, including their coherence time and spectral properties. An experimental example is given in Sect. 2.4.

In practice, various experimental imperfections including but not limited to uncontrolled polarization rotations, spectral correlation, imperfect matching of spatial modes at the beamsplitter, and timing errors mean that real photon pairs are never truly indistinguishable, and $P(1_{b_1}1_{b_2})$ does not go exactly to zero. See Sect. 2.3.1 for further discussion. The *visibility* of two-photon quantum interference is defined as

$$V = \frac{C^c - C^q}{C^c}, \tag{1.134}$$

where C^c, C^q are average coincidence count-rates $c(1_{b_1}1_{b_2})$ for the case of distinguishable (classical) and indistinguishable (quantum) input pairs, respectively. The visibility gives a useful metric of the *utility* of photons generated by a given source, and will be used throughout this thesis. When the only source of experimental imperfection is photon pair distinguishability, the visibility can be found from the density matrices of the two photons in a similar way to the purity (1.34), $V \propto Tr(\hat{\rho}_1\hat{\rho}_2)$.

For practical purposes, it would save a lot of time and money if we could reproduce the Hong-Ou-Mandel dip using attenuated laser pulses rather than expensive single-photon sources. Taking, for example, a coherent state with $\alpha = \sqrt{0.1}$,

$$|\alpha = \sqrt{0.1}\rangle = \sqrt{0.90}|0\rangle + \sqrt{0.09}|1\rangle + \sqrt{0.002}|2\rangle \ldots \tag{1.135}$$

any single-photon detection event is very likely to have originated from the $|1\rangle$ term. Naïvely, a coherent state thus appears to somehow approximate the single-photon Fock state. However, a difficulty arises in the use of many such sources—since detection is necessarily probabilistic, we cannot *synchronise* effective single-photon generation across all sources. In other words, we cannot be sure that n single-detection events corresponded to the generation of n photons in n modes, leading to temporal distinguishability and thus limited visibility of quantum interference.

Rarity et al. [49], showed that two classical beams $|\alpha\rangle_{a_1}$, $|\alpha\rangle_{a_2}$, incident on a BS with randomly varying phase, will produce a dip in coincidences as a function of temporal delay with visibility

$$V = 2\frac{\langle I_{a_1}\rangle/\langle I_{a_2}\rangle}{(\langle I_{a_1}\rangle/\langle I_{a_2}\rangle + 1)^2}, \tag{1.136}$$

where I_{a_1}, I_{a_2} are the intensities of the two input beams. For $\langle I_{a_1}\rangle = \langle I_{a_2}\rangle$, $V = 1/2$. Hence no coherent state (indeed, no classical state of light) will produce a Hong-Ou-Mandel dip with visibility $> 1/2$.

As a result, in order to see multiphoton quantum interference—which is a prerequisite for many photonic quantum technologies—we need alternative photon sources, with improved synchronicity. Ideally, we would have access to a "push-button" deterministic source of single-photon Fock states, however such devices do not currently exist. The experimental implementation of single-photon sources (SPS) providing a good approximation to this ultimate goal are discussed in Sect. 1.6.3.

1.5.3.2 Calculating States and Probabilities in Linear Optics

Throughout this thesis we will deal with circuits constructed from many linear-optical components (beamsplitters and phase-shifters) acting on a fixed, small number of photons in as many as 21 path or polarization modes. We will now outline a general method by which detection probabilities and output state vectors can be calculated for arbitrary numbers of photons, both distinguishable and indistinguishable, in generic linear-optical networks. Here we largely follow the detailed analysis of Scheel [50].

Recall that any pure superposition state can be written in the many-mode Fock basis (1.103). The input and output states for a p-photon, m-mode experiment

$$|\psi\rangle_{\text{in}} = \sum_{i=1}^{d} g_i |n_1^a, n_2^a \dots n_m^a\rangle_i; \quad |\psi\rangle_{\text{out}} = \sum_{i=1}^{d} h_i |n_1^b, n_2^b \dots n_m^b\rangle_i \tag{1.137}$$

are completely characterized by the complex probability amplitudes g_i, h_i, respectively. As before, modes a and b label the input and output ports of the circuit, although our notation has changed slightly. The states $|n_1^a, n_2^a \dots n_m^a\rangle_i$ correspond to the ith unique permutation of n photons in m modes, and together form a basis for the Hilbert space $\mathscr{H}_m^p$.

In Fock notation, we count the number of photons in each mode. Equivalently, for each photon j we can write the index z_j of the mode it occupies: For example, for two photons in three modes:

$$\begin{aligned}
&|n_1 = 2, n_2 = 0, n_3 = 0\rangle = |z_1 = 1, z_2 = 1\rangle \\
&|n_1 = 1, n_2 = 1, n_3 = 0\rangle = |z_1 = 1, z_2 = 2\rangle \\
&\quad \dots \\
&|n_1 = 0, n_2 = 0, n_3 = 2\rangle = |z_1 = 3, z_2 = 3\rangle.
\end{aligned}$$

Note that the second representation can be significantly more efficient for small numbers of photons in large circuits.

Let's consider the evolution of Fock states in an arbitrary two-mode circuit described by the matrix

$$\mathbf{\Lambda} = \begin{bmatrix} s_{11} & s_{12} \\ s_{21} & s_{22} \end{bmatrix}. \tag{1.138}$$

A single photon injected into input port a_1 evolves as

$$\mathbf{\Lambda}\, \hat{a}_{a_1}^\dagger |\mathbf{0}\rangle = \left(s_{11}\hat{a}_{b_1}^\dagger + s_{12}\hat{a}_{b_2}^\dagger\right) |\mathbf{0}\rangle = s_{11}|1_1^b 0_2^b\rangle + s_{12}|0_1^b 1_2^b\rangle. \tag{1.139}$$

Since no photons are injected into mode a_2, the second row of $\mathbf{\Lambda}$ has no effect, and the output probability amplitudes are simply $h_1 = s_{11}, h_2 = s_{12}$. When two photons injected into modes a_1 and a_2 respectively, both columns and rows of the matrix are significant:

$$\mathbf{\Lambda}\hat{a}_{a_1}^\dagger \hat{a}_{a_2}^\dagger |\mathbf{0}\rangle = \frac{1}{\sqrt{2}} s_{11}s_{21}|2_1^b 0_2^b\rangle + \frac{1}{\sqrt{2}} s_{12}s_{22}|0_1^b 2_2^b\rangle + (s_{11}s_{22} + s_{12}s_{21})|1_2^b 1_2^b\rangle. \tag{1.140}$$

For $s_{11} = s_{22} = \sqrt{t}$, $s_{12} = s_{21} = i\sqrt{r}$ we obtain the two-photon output state of a general beamsplitter, and for $r = \frac{1}{2}$ we recover the HOM dip (1.128). For the input state $|1_1 1_2\rangle$, the output probability amplitudes h_i therefore depend on $\mathbf{\Lambda}$ as

$$h_1 = \frac{1}{\sqrt{2}} s_{11}s_{21}; \quad h_2 = s_{11}s_{22} + s_{12}s_{21}; \quad h_3 = \frac{1}{\sqrt{2}} s_{12}s_{22}.$$

We can re-write these relations as

$$h_1 = \frac{1}{2\sqrt{2}} \text{per} \begin{bmatrix} s_{11} & s_{11} \\ s_{21} & s_{21} \end{bmatrix}; \quad h_2 = \text{per} \begin{bmatrix} s_{11} & s_{12} \\ s_{21} & s_{22} \end{bmatrix}; \quad h_1 = \frac{1}{2\sqrt{2}} \text{per} \begin{bmatrix} s_{12} & s_{12} \\ s_{22} & s_{22} \end{bmatrix}, \tag{1.141}$$

where $\text{per}(M)$ is the *permanent* of a matrix. The permanent of an $n \times n$ matrix M is defined in much the same way as the determinant $\det(M)$, but without the alternating sign:

$$\det(M) = \sum_{\sigma \in S_n} \text{sgn}(\sigma_i) \prod_{i=1}^{n} M_{i\sigma_i}; \quad \text{per}(M) = \sum_{\sigma \in S_n} \prod_{i=1}^{n} M_{i\sigma_i}, \tag{1.142}$$

where σ_i is a single permutation in the group S_n of all possible permutations of M.

The advantage of representing h_i in the form of (1.141) is this: Consider an arbitrary m-mode linear-optical circuit described by $m \times m$ matrices $\hat{U}$ and $\mathbf{\Lambda}$. When a given Fock state $|n_1^a n_2^a \ldots n_m^a\rangle$ is sent into the circuit, the probability amplitude h_i corresponding to detection of the state $|n_1^b n_2^b \ldots n_m^b{}_i$ at the output is in general given by

$$h_i = \langle n_1^b n_2^b \ldots n_m^b|_i \hat{U} |n_1^a n_2^a \ldots n_m^a\rangle = \left(\prod_{j=1}^{m} n_j^a! \right)^{-\frac{1}{2}} \left(\prod_{k=1}^{m} n_k^b! \right)^{-\frac{1}{2}} \text{per}\left(\mathbf{\Lambda}[z^a|z^b] \right), \tag{1.143}$$

where the expression $\mathbf{\Lambda}[z^a|z^b]$ constructs a new matrix from the columns and rows of $\mathbf{\Lambda}$, corresponding to the chosen input (z^a) and output (z^b) ports respectively [50]. To see how this works, consider again the example of $p = 2$, $m = 3$. If photons are injected at ports a_1 and a_2, the probability amplitude corresponding to coincidental detection at output ports b_2 and b_3 is proportional to the permanent of a matrix constructed from rows (1, 2) and columns (2, 3) of $\mathbf{\Lambda}$:

$$\mathbf{\Lambda} = \begin{bmatrix} s_{11} & s_{12} & s_{13} \\ s_{21} & s_{22} & s_{23} \\ s_{31} & s_{32} & s_{33} \end{bmatrix} \to \mathbf{\Lambda}[z^a|z^b] = \begin{bmatrix} s_{12} & s_{13} \\ s_{22} & s_{23} \end{bmatrix}; \quad h_i = \text{per} \begin{bmatrix} s_{12} & s_{13} \\ s_{22} & s_{23} \end{bmatrix} \tag{1.144}$$

Note that when more than one photon occupies the same mode in the input or output state, rows and columns of $\mathbf{\Lambda}$ will be repeated in $\mathbf{\Lambda}[z^a|z^b]$.

Equation (1.143) computes the probability amplitude for detection of one Fock state given another as input to the circuit. For an arbitrary superposition of input states (1.137), each output probability amplitude is simply given by a linear sum over the Hilbert space

$$h_i = \langle n_1^b n_2^b \ldots n_m^b | \hat{U} | \psi_i n \rangle = \sum_j^d \langle n_1^a n_2^a \ldots n_m^a |_i \hat{U} | n_1^a n_2^a \ldots n_m^a \rangle_j . \tag{1.145}$$

Probabilities, corresponding to experimentally detected count rates, can then be computed by the Born rule,

$$P^Q([n_1^b n_2^b \ldots n_m^b]_i) = |h_i|^2. \tag{1.146}$$

By taking the absolute-square *before* calculating the permanent, we destroy interference between different terms in $\mathbf{\Lambda}$ and hence obtain detection probabilities corresponding to distinguishable photons—classical statistics:

$$P^C([n_1^b n_2^b \ldots n_m^b]_i) = \left(\prod_{k=1}^m n_k^b! \right)^{-1} \text{per} \left(|\mathbf{\Lambda}[z^a | z^b]|^2 \right). \tag{1.147}$$

Note that the normalization constant is modified, since these are now distinguishable particles.

This method provides a very convenient route to the calculation of state vectors and detection probabilities in linear optics for arbitrary interferometers, and is used throughout this thesis. Since the technique is based almost entirely around the calculation of permanents, we can make use of the best known generic classical algorithms for per(M), rather than having to tailor our numerical methods to the physics in question.

The relationship between bosonic statistics and the permanent was first noted by Caianiello [51], and was mentioned in Valiant's 1979 proof [52] that the permanent is in general exponentially hard to compute. As such this method does not scale, and we are currently limited to problem sizes of approximately 7 photons in ~50 modes. The computational complexity of the permanent and associated linear optics experiments are discussed in depth in Sect. 6.3.2 of this thesis.

1.5.4 Interferometers

Throughout this thesis we will make use of path and polarization interferometers to manipulate and interfere quantum states of light. It will be useful to briefly examine the components and behaviour of two specific examples: the Mach-Zehnder interferometer (MZI) and the Reck-Zeilinger scheme. In Chaps. 2–5 we use MZIs to encode and manipulate qubits in a small-scale circuit model quantum processor, and the Reck-Zeilinger scheme is used in Sect. 6.3.2 to implement $m \times m$ Haar-random unitary matrices.

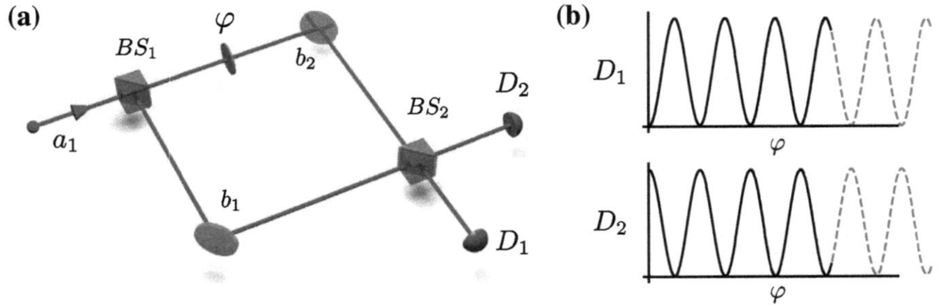

Fig. 1.6 The Mach-Zehnder interferometer. **a** A light beam is divided into two paths by a beamsplitter, and one path is phase-shifted with respect to the other by by a relative phase φ. The two beams are then mixed on a second beamsplitter, giving rise to **b** interference fringes in the measured intensity at detectors D_1, D_2

1.5.4.1 The Mach-Zehnder Interferometer

A typical bulk-optical MZI is shown in Fig. 1.6. The MZI has two input ports corresponding to (a_1, a_2) of a 50:50 beamsplitter, which splits a beam injected into either port into two paths. A relative phase-shift φ, equivalent to a path-length difference $dz = \varphi\lambda/2\pi$, is introduced into one arm. The two paths are then mixed at a second 50:50 BS, the output ports of which are monitored by single-photon detectors or photodiodes, D_0 and D_1. A phase shift acting, for example, on arm b_2 of the interferometer transforms $\hat{a}^\dagger_{b_2} \rightarrow e^{i\varphi}\hat{a}^\dagger_{b_2}$ and can be written as a unitary matrix

$$\hat{U}_{\mathrm{PH}}(\varphi) = \begin{bmatrix} 1 & 0 \\ 0 & e^{i\varphi} \end{bmatrix} = e^{-i\varphi/2}\begin{bmatrix} e^{i\varphi/2} & 0 \\ 0 & e^{-i\varphi/2} \end{bmatrix} \rightarrow \begin{bmatrix} e^{i\varphi/2} & 0 \\ 0 & e^{-i\varphi/2} \end{bmatrix}, \quad (1.148)$$

where we have chosen to ignore the global phase $-\varphi/2$ as it cannot be measured in the two-mode system considered here. We can then write the matrix corresponding to the entire MZI,

$$\hat{U}_{\mathrm{MZI}}(\varphi) = \hat{U}_{BS_2}\hat{U}_{\mathrm{PH}}(\varphi)\hat{U}_{BS_1} \quad (1.149)$$

$$= \frac{1}{\sqrt{2}}\begin{bmatrix} 1 & i \\ i & 1 \end{bmatrix}\begin{bmatrix} e^{i\varphi/2} & 0 \\ 0 & e^{-i\varphi/2} \end{bmatrix}\frac{1}{\sqrt{2}}\begin{bmatrix} 1 & i \\ i & 1 \end{bmatrix} = i\begin{bmatrix} \sin(\varphi/2) & \cos(\varphi/2) \\ \cos(\varphi/2) & -\sin(\varphi/2) \end{bmatrix}, \quad (1.150)$$

where the global phase i can again be neglected. If light is injected into port a_1 of the MZI, the intensity at (D_1, D_2) therefore depends on the phase φ as

$$I_{D_1} = I_0 \sin^2(\varphi/2); \quad I_{D_2} = I_1 \cos^2(\varphi/2). \quad (1.151)$$

These are the interference fringes as shown in Fig. 1.6b.

Note that $\hat{U}_{\text{MZI}}$ strongly resembles a variable-reflectivity beamsplitter (1.119) $\hat{U}_{\text{BS}}(r)$. In fact, by applying phase shifts before and after the MZI the circuit can be made identical to a beamsplitter with arbitrary reflectivity r:

$$\hat{U}_{\text{BS}} = \begin{bmatrix} \sqrt{t} & i\sqrt{r} \\ i\sqrt{r} & \sqrt{t} \end{bmatrix} = \begin{bmatrix} \sin(\varphi/2) & i\cos(\varphi/2) \\ i\cos(\varphi/2) & \sin(\varphi/2) \end{bmatrix} = -i \begin{bmatrix} 1 & 0 \\ 0 & i \end{bmatrix} \hat{U}_{\text{MZI}}(\varphi) \begin{bmatrix} 1 & 0 \\ 0 & i \end{bmatrix}, \tag{1.152}$$

where $\varphi = 2\cos^{-1}(\sqrt{r})$. The MZI structure therefore allows us to convert a passive device with fixed beamsplitter reflectivities into a *reconfigurable* device by adding controlled phase-shifts. Experimentally it is often considerably easier to dynamically control a phase shift than a beamsplitter reflectivity, and this technique is used extensively throughout this thesis.

We can further extend this result to show that an MZI with external phase shifts can implement *any unitary operator* in the group $SU(2)$, i.e. any lossless two-mode operation. To see this, note that $\hat{U}_{\text{MZI}}$ with external phaseshifts (1.152) is identical to rotation by an angle $\theta = \varphi + \pi$ about the x-axis of the Bloch sphere (Fig. 1.2),

$$\hat{R}_x(\theta) = e^{i\theta\hat{\sigma}_x/2} = -i \begin{bmatrix} 1 & 0 \\ 0 & i \end{bmatrix} \hat{U}_{\text{MZI}}(\varphi) \begin{bmatrix} 1 & 0 \\ 0 & i \end{bmatrix}, \tag{1.153}$$

and that a phase shifter (1.148) corresponds to a rotation about the z-axis, $\hat{U}_{\text{PH}}(\varphi) = \hat{R}_z(\varphi) = e^{i\varphi\sigma_z/2}$. A simple geometric argument leads to the observation that these two rotations are sufficient to take any point on the Bloch sphere to any other point, including a global phase—that is, to map any pure state to any other pure state of a two-level system. Any unitary operator in $SU(2)$ can therefore be realised using an MZI with phaseshifters at the input and output:

$$\hat{U} = \hat{R}_z\gamma\hat{R}_x\beta\hat{R}_z\alpha = e^{i\gamma\hat{\sigma}_z}e^{i\beta\hat{\sigma}_x}e^{i\alpha\hat{\sigma}_z} = \hat{U}_{\text{PH}}(\gamma')\hat{U}_{\text{MZI}}(\beta')\hat{U}_{\text{PH}}(\alpha'), \tag{1.154}$$

where α, β, γ are real numbers—the Euler angles [53]. A single photon in an MZI thus provides a convenient encoding for the two level system of a *qubit*—this is discussed in further detail in Sect. 1.6.1.1.

Note that the interferometer is sensitive to phase shifts on the order of the wavelength $dz \sim \lambda$. This sensitivity allows the MZI to be used for extremely precise interferometric measurements of distance, refractive index, and other optical properties of interest. However, this sensitivity is a double-edged sword—in order to construct a stable MZI we must ensure that the relative positions of the beamsplitters and mirrors are static to within a small fraction of the optical wavelength, i.e. ~nm. In a bulk optical setup, this is extremely difficult to achieve due to thermal expansion/contraction and acoustic vibration of the apparatus. Although intrinsically stable bulk optical interferometers can be built, for instance using beam displacers [54] or a Sagnac architecture [55], these schemes introduce further complexity and are not scalable. As a result, many experiments in quantum optics use polarization encoding in free-space, which is intrinsically stable—as only a single spatial mode is used.

With the recent advent of integrated quantum photonics (IQP) it has become possible to build complex, multi-mode interferometers on-chip. By embedding the interferometer in a monolithic substrate, stable path-interferometry can be scaled to devices with thousands of optical modes, while simultaneously being miniaturized by a factor of a million [56] with respect to equivalent bulk-optical apparatus. Path-encoding then becomes a very natural choice, particularly since on-chip polarization-encoding is currently problematic. This topic is discussed in detail in Sect. 1.6.5.

1.5.4.2 Linear-Optical Implementation of Any Unitary Operator

As we have already seen, an MZI surrounded by two phase-shifters can implement an arbitrary unitary operation on two modes ($\hat{U} \in SU(2)$). How does this generalize to circuits with more than two modes? What is the class of operations that we can implement on m modes, using only linear-optical elements?

As shown by Reck and Zeilinger [57], *any* $m \times m$ *unitary operator* $\hat{U}$ corresponds to a linear-optical circuit on m modes, constructed from beamsplitters and phase-shifters only. That is, any $\hat{U}$ has a decomposition as a product

$$\hat{U} = \hat{U}_T \cdot \hat{U}_{T-1} \dots \hat{U}_1, \tag{1.155}$$

where each $\hat{U}_T$ acts nontrivially on at most two modes and does not affect the remaining m-2 modes. A simple proof has been given by Aaronson and Arkhipov [58]. We have already seen that each two-mode $\hat{U}_T$ can always be implemented using an MZI with a total of three phase shifters. Given a target unitary $\hat{U}$, the task is then to perform the decomposition (1.155). In fact, this decomposition is equivalent to a standard technique for QR decomposition[6] of a matrix using *Givens rotations*—2×2 matrices corresponding to $\hat{U}_T$. The circuit for $\hat{U}$ in terms of optical elements $\hat{U}_T$ can thus be found for any discrete $\hat{U}$.

In their paper, Reck and Zeilinger go on to show how this decomposition can be implemented using a *single* linear optical network, which is reconfigured by means of phase shifters and variable beamsplitters to implement any $\hat{U}$. In general, the circuit uses $O(m^2)$ elements. In this design, the network is *local* in the sense that each $\hat{U}_T$ acts on pairs of adjacent waveguides, considerably simplifying the experimental implementation. The general form of the circuit is shown in Fig. 1.7a. The design lends itself to an implementation in using integrated optics, where interferometric stability is simple to achieve. Equivalent waveguide circuits are shown in Fig. 1.7b, d.

Although the scheme is perhaps more easily visualized in path, it should be emphasised that the modes m can in principle correspond to any degree of freedom of the photon, so long as the corresponding beamsplitter and phase-shifter operations can

[6]This implies that numerical methods identical to the Reck-Zeilinger decomposition are provided in almost any numerical linear-algebra package capable of QR decomposition (e.g. LAPACK). Your home router probably knowns how to build Reck schemes.

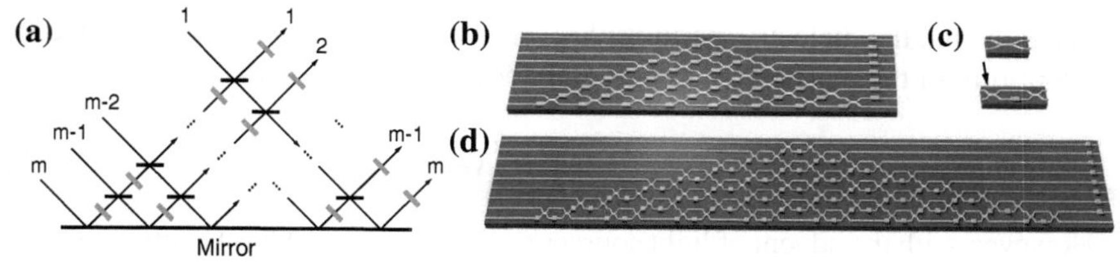

Fig. 1.7 Any unitary operator $\hat{U}$ on m optical modes can be implemented using 2×2 optical elements—beamsplitters and phase-shifters. **a** Original figure, reproduced from [57]. Blue lines represent phase-shifters, short black lines are beamsplitters with arbitrary reflectivity. **b** Stabilization of the interferometer in **a** would be practically very challenging in a bulk-optical architecture. The same circuit can instead be implemented using integrated photonics (Sect. 1.6), providing interferometric stability and miniaturization. This circuit is implemented, without phase-shifters, in Sect. 6.3.2 of this thesis. **c** Beamsplitters drawn in **a, b** must have variable reflectivity. In an integrated circuit, we replace each variable beamsplitter by an MZI, allowing the effective reflectivity of each splitter to be controlled by a phase-shifter, leading to the circuit shown in **d**

be constructed. A recent example [59] uses a combination of path and polarization modes in a bulk-optical setup.

If we can use Reck-Zeilinger to implement any unitary matrix, does that mean that we can build a universal gate-set for a quantum computer using only beamsplitters and phase-shifters? To answer this question, it should be emphasised that Reck-Zeilinger allows us implement an arbitrary unitary on *modes*, whereas $\hat{U}_{\text{CNOT}}$ acts on *qubits*. Using a Reck scheme we can implement any $m \times m$ matrix dictating the dynamics of a *single* photon in an m-mode circuit, i.e. acting on the single-photon Hilbert space $\mathscr{H}_m^1$. Following the method outlined in Sect. 1.5.3.2, this then *generates* the $d \times d$ matrix $\mathcal{U}$ acting on the full Hilbert space of p photons in m modes, $\mathscr{H}_m^p$, which is in general exponentially larger ($d = \binom{m+1-p}{p}$). Since this is the space onto which our qubits are mapped, by a simple parameter-counting argument we cannot always use Reck-Zeilinger to deterministically implement arbitrary unitary operations on photonic qubits. In principle, we could map n qubits to the state of a single photon in 2^n modes, in which case $\hat{U} = \mathcal{U}$ and Reck-Zeilinger *can* be used to implement universal quantum computing—but the necessary experimental resources clearly scale exponentially in n. The latter scheme, which cannot provide an exponential speedup over classical machines, has recently been suggested for superconducting qubits [60].

1.5.5 Nonlinear Optics

The majority of optical effects observed in nature are linear, in the sense that the properties of the material or medium are independent of the incident light field. Under these conditions, the wavelength of light is not changed when passing through the medium, and a light source will never have control over the behaviour of another.

In linear media, the dielectric perimittivity ε is a constant function of the dielectric susceptibility of the material χ_e (1.60), and does not depend on the electric field

$$\varepsilon(\mathbf{E}) = \varepsilon_0 [1 + \chi_e]. \tag{1.156}$$

However, with the advent of light sources such as the laser, it has become possible to engineer situations in which the passage of an intense light beam through an optical medium temporarily modifies the properties of the material itself to a significant extent. This can generate new optical fields or cause self-modulation of the incident beam, allowing "light to control light" where that control is mediated by the optical material.

The effect of a strong optical field incident on a nonlinear medium can be described in terms of the dielectric polarization vector **P**, which is introduced into the expression for the electric flux density **D** in Gauss' law (1.59) as

$$\mathbf{D}(\mathbf{E}) = \varepsilon\mathbf{E} = \varepsilon_0 [1 + \chi_e] \mathbf{E} \quad \rightarrow \quad \mathbf{D}(\mathbf{E}) = \varepsilon_0 \mathbf{E} + \mathbf{P}(\mathbf{E}) \tag{1.157}$$

where

$$\mathbf{P}(\mathbf{E}) = \varepsilon_0 (\chi_e^{(1)} \mathbf{E} + \chi_e^{(2)} \mathbf{E}^2 + \chi_e^{(3)} \mathbf{E}^3 + \cdots) \tag{1.158}$$

Here $\chi_e \equiv \chi_e^{(1)}$ is the standard (linear) dielectric susceptibility, while $\chi_e^{(2)}$ etc. characterise the higher-order nonlinear response of the material. In most nonlinear media the magnitude of these terms decreases rapidly with order, i.e.

$$\chi_e^{(1)} \gg \chi_e^{(2)} \gg \chi_e^{(3)} \ldots \tag{1.159}$$

and in order for $\chi_e^{(2)}$ to be nonzero, the material must be birefringent.

This nonlinear response allows nonlinear materials to mediate an effective interaction between photon pairs. However, since $\chi_e^{(1)} \gg \chi_e^{(2)}$, any such effect is typically very weak. As a result it is technically very difficult to use such media to entangle two photons initially prepared in a separable state, for example. This difficulty, together with a potential solution to effective photon interaction which does not directly depend on intrinsic optical nonlinearity, is discussed further in Sect. 1.6.2.

1.6 Quantum Photonics

In order to implement any of the quantum technologies described in Sect. 1.4, we must first choose a physical system in which to encode quantum information. As already discussed, this system should support the preparation, controlled coherent manipulation, and readout of single quanta. This leads to a challenging set of near-incompatible requirements: In order to avoid decoherence and the unwanted introduction of mixture, the system must be carefully protected from interaction with the environment, while, at the same time—in order to achieve the entangling operations required for most quantum technologies—amenable to strong, controlled pairwise interaction. Moreover, the experimentalist should have access to a number of control parameters with direct influence on the system's state.

Over the past few decades, a range of physical systems have emerged as leading solutions to this problem. Cold atoms [61] and charged ions [62], held in a variety of electromagnetic traps, satisfy many of the desired criteria, in particular the availability of strong pairwise interaction. Superconducting qubits [63], based on Josephson junctions, as well as nitrogen vacancy (NV) centers in diamond [64] and phosphorous impurities in silicon [65], are more immediately amenable to monolithic integration, and have recently seen considerable industrial interest [66]. However, ions, atoms, and spins all readily interact with both light and matter and the major limiting factor of many of these matter-based platforms is environment-induced decoherence. Much of the experimental challenge therefore involves the careful isolation of the system of interest from environmental effects, often requiring ultrahigh vacuum and/or cryogenic temperatures.

These difficulties lend favour to the prospect of an all-optical photonic quantum computer, where qubits are encoded in the quantum state of single photons. In general, photons interact only very weakly with their environment, and single photons propagating in free space or optical fibre at room temperature and pressure (RTP) suffer negligible decoherence. Over the past half-century, single photon sources (Sect. 1.6.3), and high-efficiency single photon detectors (Sect. 1.6.4) have become widely available. Deterministic single-qubit operations are very easily implemented using passive linear optics, as described in Sect. 1.5.4. Many classical imaging and measurement techniques are optical, and photons are a natural choice for many applications of quantum metrology. Owing to their speed, photons are also natural candidates when quantum information must be moved over an appreciable distance, either between registers in a quantum computer, or over long-distance communication channels [67].

High-fidelity quantum states of single photons are now routinely generated, manipulated and measured at RTP, and many early demonstrations of quantum effects, including superposition [68], nonlocality [15], large-scale entanglement [69], two-qubit gates [55], QKD [70], quantum metrology [71], quantum algorithms [72–74], ECCs [75], etc. have used single photons at near-visible wavelengths.

1.6.1 Photons as Qubits

A single photon is associated with a number of continuous variables, including position and frequency, and in general occupies an infinite-dimensional Hilbert space. In order to encode a photonic qubit, we must therefore restrict the dynamics to an effective two-level system. In principle this can be achieved using a single cavity mode, mapping logical qubit states $|0\rangle$ and $|1\rangle$ to the vacuum and single-photon

Fock state respectively. However, this simple encoding has obvious drawbacks: for instance, rotation of a single qubit from $|0\rangle$ to $|1\rangle$ becomes experimentally challenging, requiring a photon source.

Instead, it is experimentally much more convenient to use two modes and one photon per qubit. Modes in frequency, time, and orbital angular momentum [76] are routinely used to encode quantum information, however, in this thesis we will only consider *path encoding* and *polarization encoding*.

1.6.1.1 Path Encoding

Path encoding, otherwise known as *dual-rail* encoding, stores a qubit as a propagating photon in a superposition of two optical spatial modes a_0 and a_1. The two logical-basis states of the qubit, $|0\rangle$ and $|1\rangle$, correspond to states of the photon occupying each spatial mode respectively. Mapping from qubits to the Fock-state representation,

$$\alpha|0\rangle + \beta|1\rangle \equiv \alpha|1_{a_0}0_{a_1}\rangle + \beta|0_{a_0}1_{a_1}\rangle. \tag{1.160}$$

As described in Sect. 1.5.4, deterministic, arbitrary unitary operations on two spatial modes are easily accomplished using an MZI. Any path-encoded state can thus be mapped to another using beamsplitters and phaseshifters. Techniques for state preparation and measurement of path-encoded qubits are shown in Sects. 2.25 and 2.26.

Path encoding has the advantage of easily scaling to higher-dimensional *qudit* encodings, where a d-level system is encoded using a single photon together with d spatial modes. The result of Reck-Zeilinger (Sect. 1.5.4.2) allows arbitrary deterministic rotations of path-encoded qudits using beamsplitters and phaseshifters only. This possibility is discussed further in Sect. 6.3.

As long as we can engineer single-mode optics, path-encoding is relatively easy to implement. However, when realised using bulk optics, thermal instability and mechanical vibration of the experimental setup will give rise to uncontrolled time-varying phase shifts in the interferometer. This has the effect of adding mixture to the state, and is largely indistinguishable from decoherence. Although active stabilization or Sagnac architectures can be used to overcome this difficulty, these techniques are expensive and complicated and, for bulk optical setups, path-encoding has largely been avoided in favour of polarization-encoded qubits.

More recently, IQP (Sect. 1.6.5), which provides inherent interferometric stability, has enabled path-encoding on a large scale.

1.6.1.2 Polarization Encoding

Path-encoding suffers from the difficulty of nm path-length matching, and as such is very challenging to implement in bulk-optics, or when communicating over long distances. *Polarization encoding*, in which the logical basis states of the qubit are mapped to the horizontal $|H\rangle$ and vertical $|V\rangle$ polarization states of the photon, overcomes these problems. Since both polarizations propagate in the same spatial mode, there is no difficulty of path-length matching. Deterministic arbitrary single-qubit rotations on polarization-encoded qubits can easily be accomplished using a system of birefringent quarter-wave and half-wave plates, following a decomposition of $\hat{U}$ which is analogous to that of the MZI. Polarization-encoding has the further advantage that polarization-entangled states are naturally generated by spontaneous parametric downconversion (SPDC), as described in Sect. 1.6.3.1.

Polarization encoding is not amenable to qudit encodings. Moreover, the ability to faithfully transport and manipulate polarization-encoded states in optical waveguides is not currently well-developed, as described in Sect. 2.2.1.

In general, we can deterministically convert between path and polarization encodings using a polarising beamsplitter (PBS), which transits and reflects horizontally and vertically polarized light, respectively. Using a similar notation to Fig. 1.5,

$$\hat{U}_{\mathrm{PBS}} = |H_{a_1}\rangle\langle H_{b_1}| + |H_{a_2}\rangle\langle H_{b_2}| + i|V_{a_1}\rangle\langle V_{b_2}| + i|V_{a_2}\rangle\langle V_{b_1}|. \tag{1.161}$$

1.6.2 Linear-Optical Quantum Computing

In Sect. 1.6, we argued that photonics offers an advantage over many other approaches to the implementation of quantum technologies, owing to the inherent reluctance of photons to interact with their environment. However, this comes at a cost, in that photons are *also* very reluctant to interact with one another. This presents a serious challenge to the implementation of entangling operations required by many quantum technologies. Direct photon-photon interaction is so weak as to never be seen outside a particle accelerator. Although nonlinear Kerr media (Sect. 1.5.5) can be used to mediate an effective interaction between photons, this effect is many orders of magnitude too weak ($\chi^{(3)} \approx 1 \times 1^{-22}\,\mathrm{m^2\,V^{-2}}$) to be feasible. Extremely strong optical non-linearities can be obtained when photons interact with a solid-state atom-like system, such as a charged ion or a quantum dot, however, the current performance of these technologies, particularly with respect to loss and coupling strength, is far from sufficient for quantum computation [77, 78].

As a result, it may then appear that photonic quantum computing is forbidden by strong technological constraints. In 2001, Knill, Laflamme and Milburn (KLM) set out to formalize this reasoning, in order to show that without a strongly nonlinear optical medium or component, scalable photonic quantum computing should be impossible. To the surprise of many, they found [79] the converse: that full-scale, universal quantum computation can be scalably achieved using only single-photon sources, single photon detectors, and a linear-optical network, together with *adaptive measurement*, a.k.a. feed-forward.

At the heart of the Knill, Laflamme and Milburn (KLM) quantum computer is HOM interference, as described in Sect. 1.5.3. As has already been discussed, indistinguishable bosons in linear-optical circuits exhibit highly non-classical interference effects, and generate correlations which cannot be classically reproduced. However, as was shown by Kok and Braunstein [80], these phenomena cannot be used to implement *deterministic* entangling gates on photonic qubits. For example, the 2-photon NOON state (1.128) generated by a BS is entangled, but it is not obvious how to convert this state to a Bell state (1.38) using linear optics alone. The first insight of KLM was to show that quantum interference of Fock states in a simple linear-optical network could *probabilistically* implement a maximally entangling operation on two qubits. A construction and experimental implementation of a two-qubit gate derived from the original proposal of KLM is given in Sect. 2.2.4.

A fundamentally probabilistic gate is problematic for scalable quantum computation, as the success probability of composite circuits built from such gates will in general fall off exponentially with circuit size. The second, extremely significant result of KLM was to show that such probabilistic gates can be *bootstrapped* into a scalable architecture, using ancillary photons together with measurement and feed-forward. Sending extra photons into the circuit, which are not used to encode logical qubits, detection events registered at the output can then be used to obtain classical information on the success or failure of the gate. This information is then used to reconfigure the circuit downstream of the gate, essentially correcting for failure. KLM showed that this feed-forward technique can be used to render linear-optical entangling gates asymptotically deterministic, with only a polynomial resource overhead. Specifically, KLM give a linear-optical construction for a maximally entangling controlled-Z (CZ) gate with success probability scaling as $p^2/(p+1)^2$ in the number of ancilla photons p.

By removing the need for strong natural optical non-linearities, the result of KLM significantly reduces the experimental difficulty of photonic quantum computation, and as a result has attracted considerable experimental interest [81, 82]. However, the resource overhead necessary for scalable operation, while polynomial, is prohibitively large for real-world implementations. Fortunately, a number of recent proposals [83, 84] have significantly improved on the original result. Using a *one-way* model of quantum computation based on the generation and measurement of *cluster states*, these schemes dramatically reduce the resource overhead required for scalability, to the extent that realization of linear-optical quantum computation is now arguably more of an engineering challenge than an open theoretical question. A number of experimental implementations have since been reported [85–87].

1.6.3 Sources

We have already seen that the coherent state generated by a laser (Sect. 1.5.2) is not appropriate for experiments which depend on multiphoton quantum interference. Most photonic quantum technologies depend on light sources which do not admit a classical description. Arguably the most technically demanding is the *on-demand single-photon source*. This would be a device which deterministically generates indistinguishable single-photon Fock states $|0\rangle$ in a single mode, on demand. Currently, no such device exists, and the development of scalable SPSs remains a very significant challenge for the realization of quantum technologies. A scalable on-demand SPS would have immediate applications for QKD [88], and metrology [38], and would represent a very significant step towards tangible quantum speedup in information processing tasks (Sect. 6.3.2).

Leading candidates for deterministic single-photon sources include artificial-atom systems [89] such as NV centres in diamond, quantum dots, and various atomic systems. There is no fundamental limit to the probability of success of such SPSs. However, these techniques currently do not achieve sufficient performance—in particular, with respect to out-coupling efficiency and photon indistinguishability—to be immediately applicable to the demanding multiphoton experiments described in this thesis.

Historically, a great many proof-of-principle demonstrations of quantum information tasks have been accomplished using non-deterministic SPSs based on parametric nonlinear optical processes. We will focus our discussion on these sources, which are used throughout the experiments described in this thesis.

Non-deterministic, spontaneous photon sources do not directly provide a route to scalable quantum technologies, as the probability of generating p indistinguishable photons falls off exponentially with p. However, it has recently been suggested [90, 91] that by *multiplexing* many nondeterministic sources in parallel, together with single-photon detection and a fast switching network, it should be possible to construct an asymptotically deterministic on-demand source with polynomial resource overhead. This provides an alternative route to a scalable single-photon source, which is particularly amenable to monolithic integration (Sect. 1.6.5).

1.6.3.1 Spontaneous Parametric Down-Conversion

Nonlinear optics (Sect. 1.5.5), when combined with single photon detection (Sect. 1.6.4), provides a convenient and historically very successful route to approximate, non-deterministic single-photon sources.

The result of the $\chi^{(2)}$ nonlinearity introduced in (1.158) is to allow so-called *3-wave mixing* effects. These include *sum-frequency generation* in which two pump beams with frequencies (ω_1, ω_2) generate a new optical field with $\omega_1 \pm \omega_2$, and spontaneous parametric downconversion (SPDC), in which a single pump beam ω_0 generates two daughter fields with frequencies ω_1 and ω_2. SPDC allows a light beam to be arbitrarily down-converted to a longer wavelength, and as such has many classical applications. In this thesis we are principally concerned with SPDC as a source of quantum states of light—single photons.

In the quantum picture of SPDC, a high-energy pump photon in a single mode with wavevector $\mathbf{k}_0$ is incident on a nonlinear birefringent crystal with a $\chi^{(2)}$ nonlinearity. The pump photon splits into two daughter photons in modes $\mathbf{k}_1$, $\mathbf{k}_2$, referred to as the *signal* and *idler* for historical reasons. This process must of course preserve conservation of energy and momentum, having

$$\omega_1 + \omega_2 = \omega_0; \qquad \mathbf{k}_1 + \mathbf{k}_2 = \mathbf{k}_0. \tag{1.162}$$

Throughout this thesis we will optimize our sources to generate indistinguishable photon pairs with $\omega_1 = \omega_2$.

Adding the interaction terms generated by (1.158) to the quantized Hamiltonian of the free electromagnetic field (1.97) and summing over all modes, we can write the SPDC Hamiltonian [92]

$$\hat{H} = \sum_{i=0}^{2} \hbar\omega_i \left(\hat{n}_i + \frac{1}{2}\right) + \hbar g \left[\hat{a}_1^\dagger \hat{a}_2^\dagger \hat{a}_0 + \text{h.c.}\right], \tag{1.163}$$

where $\hat{a}_1^\dagger$, $\hat{a}_2^\dagger$ are creation operators for photons in the signal and idler modes respectively, $\hat{a}_0$ corresponds to annihilation of the pump photon, and $g \propto \chi^{(2)}$ is a coupling constant which ensures that the conditions of (1.162) are met.

Usually the applied pump is an intense laser beam, modelled by the coherent state $|\alpha\rangle$ with $\langle\hat{n}_1(t)\rangle, \langle\hat{n}_2(t)\rangle \ll |\alpha|^2$. Since the pump field is then effectively classical, we can re-write the interacting part of $\hat{H}$ as

$$\hat{H}_I = i\xi\hbar \left(\hat{a}_1^\dagger \hat{a}_2^\dagger + \text{h. c.}\right), \tag{1.164}$$

where the classical properties of the pump, including the fast modulation $e^{-i\omega_0 t}$, have been lumped together with g into ξ. Assuming that the signal and idler modes are initially prepared in the vacuum state $|0_1 0_2\rangle$, time evolution of the system is then governed by the unitary operator $\hat{U} = e^{-i\hat{H}_I t/\hbar}$, leading to an output state

$$\begin{aligned}
|\Psi_{\text{SPDC}}\rangle &= \hat{U}|0_1 0_2\rangle & (1.165)\\
&\approx e^{\xi\hbar t \hat{a}_1^\dagger \hat{a}_2^\dagger}|0_1 0_2\rangle & (1.166)\\
&= \sum_{j=0}^{\infty} \frac{\gamma^j}{j!} \left(\hat{a}_1^\dagger\right)^j \left(\hat{a}_2^\dagger\right)^j |0_1 0_2\rangle = \sum_{j=0}^{\infty} \gamma^j |j_1 j_2\rangle & (1.167)\\
&= |0_1 0_2\rangle + \gamma|1_1 1_2\rangle + \gamma^2|2_1 2_2\rangle + \gamma^3|3_1 3_2\rangle \ldots & (1.168)
\end{aligned}$$

where $\gamma = t\xi$ and we have assumed that $|\gamma| \ll 1$.

The importance of the SPDC state (1.168) for applications in quantum photonics is this: when γ is small such that $\gamma \gg \gamma^2 \gg \gamma^3 \ldots$, the state $|\Psi_{\text{SPDC}}\rangle$ is well-approximated by a superposition of the vacuum and a two-photon state $|1_1 1_2\rangle$:

$$|\Psi_{\text{SPDC}}\rangle \approx |\mathbf{0}\rangle + \gamma|1_1 1_2\rangle. \tag{1.169}$$

A single-photon detection event in the idler arm therefore *heralds* a single-photon Fock state in the signal arm with high probability, and vice-versa. Moreover, by using two detectors and counting in the *coincidence basis* (i.e. only registering events in which both signal and idler detectors clicked) we post-select on the $|1_1 1_2\rangle$ term, allowing $|\Psi_{\text{SPDC}}\rangle$ to be used as an approximate source of indistinguishable photon pairs.

Most experiments in quantum optics are performed using non-number resolving ("bucket") detectors, which cannot distinguish between Fock states $|n\rangle$. If a coincidence click is registered across the signal and idler modes, there is a small probability $|\gamma|^4$ that this event came from the $|2_1 2_2\rangle$ term in (1.168), leading to partial mixture of the effective experimental state. Increased pump power, while increasing the overall downconversion rate, leads to a greater value of γ and an increased relative probability of detection events due to higher-order terms. If the desired state is $|1_1 1_2\rangle$, as is the case throughout this thesis, this effect degrades the quality of the measured state.

The allowed signal and idler modes are those which meet the conditions energy conservation and phase-matching (1.162). This depends on the experimental geometry, the nonlinear material, the pump, signal and idler wavelengths, and a variety of other experimental parameters. In *type-I phase-matching*, photon pairs with identical polarization are generated at diametrically opposed points on a cone centred about the pump axis (Fig. 1.8a). The opening angle of the cone depends on the pump wavelength and the properties of nonlinear material—in particular, the orientation of the crystal lattice with respect to the pump beam. Photons generated by type-I SPDC are entangled in wavelength, time, and space, but not in polarization, and we therefore collect the state $|V_1 V_2\rangle$. In *type-II phase-matching*, photons are generated in *two* overlapping cones with orthogonal polarization [93] as illustrated in Fig. 1.8b. At the points where the cones overlap, since we cannot distinguish one photon from another nor from which cone either photon was collected, the state is entangled in polarization across four modes (paths 1, 2 and polarizations H, V),

$$|\Psi_{\text{SPDC-II}}\rangle \propto \sum_{n=0}^{\infty} \gamma^n \left[\sum_{m=0}^{n} (-1)^m |n - m_{H1}, m_{V1}, m_{H2}, n - m_{V2}\rangle \right] \tag{1.170}$$

$$= |\mathbf{0}\rangle + \gamma |1_{H1}, 0_{V1}, 0_{H2}, 1_{V2}\rangle - \gamma |0_{H1}, 1_{V1}, 1_{H2}, 0_{V2}\rangle + \text{h.f.} \tag{1.171}$$

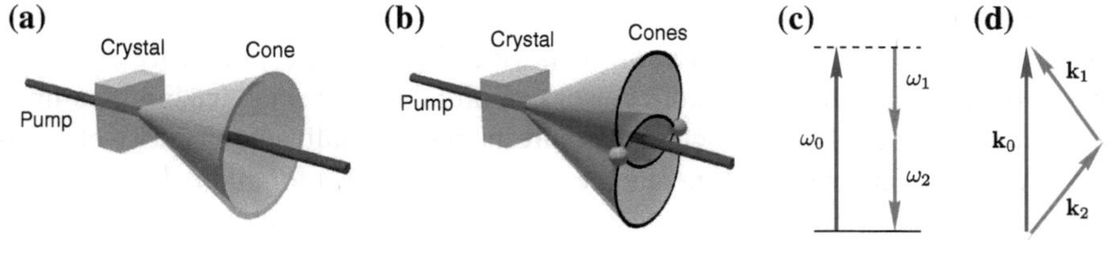

Fig. 1.8 **a** Type-I SPDC cone structure. Downconverted photon pairs are generated at diametrically oppposed points about the pump axis, on a cone with a typical opening angle of ~3°. **b** Type-II cones. Entangled photon pairs lie at the intersection of the two cones (*green spheres*). **c** Conservation of energy. **d** Conservation of momentum: the phase-matching condition (color online)

After post-selection on detection of one photon in each spatial mode and re-normalization this is equivalent to

$$|\Psi^-\rangle = \frac{1}{\sqrt{2}} \left(|H_1 V_2\rangle - |V_1 H_2\rangle \right), \tag{1.172}$$

which is a maximally entangled Bell state (Sect. 1.3.7).

Most of the experimental work in this thesis makes use of type-I SPDC to generate indistinguishable photon pairs, both in the CW and pulsed regimes. The exception is Sect. 4.5, in which type-II SPDC is used to generate polarization-entangled states in the form of (1.172).

1.6.4 Detectors

In order to read-out quantum information from a photonic system, we must almost always use single-photon detectors. Classical detectors, sensitive only to macroscopic light intensity, are usually not sufficient to obtain a quantum advantage. When a single photon, (typically with energy $\hbar\omega \approx 10 \times 10^{-21}$ J) is incident on the active area of a single-photon detector, we would like to raise a macroscopic, classically accessible flag or signal. Ideally, this process would be deterministic and fast, allowing detection events to be correlated in time. Such an idealised single-photon detector, acting on a mode k, is described in the Fock basis by the projector $\Pi_d = |1_k\rangle\langle 1_k|$, with

$$\mathrm{Tr}\left(\Pi_d |\mathbf{0}\rangle\langle\mathbf{0}|\right) = 0; \quad \mathrm{Tr}\left(\Pi_d |1\rangle\langle 1|\right) = 1. \tag{1.173}$$

All practical single-photon detectors face the difficulty of amplifying the small change in energy imparted by a single photon to the macroscopic level. As a result, real-world single-photon detectors suffer from a number of imperfections, the most significant of which is limited detection (quantum) efficiency. Strong amplification also leads to electrical noise, which manifests as so-called *dark counts*—signals which positively indicate single-photon detection, when no photon was incident on the detector. Moreover, all electronic signals suffer from timing uncertainty or *jitter*, limiting the timing resolution of the device. The amplification process is often based on an avalanche or breakdown from an initial fragile state, leading to a finite *dead-time*, during which the detector is unresponsive. Finally, the majority of existing single-photon detectors generate the same output signal for all Fock states other than the vacuum—that is, they are not sensitive to the photon *number*. In Sect. 6.3.3, we experimentally test pseudo-number resolving detectors constructed from many non-number-resolving parts.

The detectors used throughout this thesis were *Perkin-Elmer* silicon avalanche photodiodes (APDs), operating in Geiger (free-running) mode. A strong reverse bias is applied to a silicon P-N junction, such that a single incident photon is sufficient to raise an electron from the valence band into the conducting band, triggering an avalanche of electric current amplification, and leading to a voltage pulse across the diode. This pulse is detected and conditioned by a digital microprocessor, which ultimately outputs a clean TTL pulse for time-correlated counting. Silicon APDs typically achieve a quantum efficiency of $\sim$60 % at 808 nm, although this can vary significantly between devices, and exhibit typical dark-count rates on the order of 100 Hz. While the diode itself is maintained significantly below room temperature by a Peltier cooling system, Si APDs do not require cryogenic cooling, facilitating our experiments.

1.6.5 Integrated Quantum Photonics

Bulk optics has historically been very successful as a platform for proof-of-principle tests of quantum physics, as well as rapid prototyping of quantum technologies. However, this approach—in which cm-scale optical elements are bolted to a $\sim$3 m $\times$ 1.5 m optical bench weighing $\sim$1 t—is not expected to scale to experiments demanding large numbers of photons or qubits. First, there is simply not enough physical space in a typical laboratory. Secondly, as the complexity of the optical apparatus is increased, the demand on the experimentalist in terms of alignment and stabilization grows rapidly.

In recent years, as optical networking, energy efficiency, and parallelism have become increasingly important for general-purpose computing, there has been renewed interest in the all-optical transport, switching, and processing of large volumes of classical information. In the course of development of these technologies, which include optical interconnects and fast fiber-optic network switches, there has been considerable investment in the field of *integrated photonics*: monolithic, miniaturized chips which generate, guide, manipulate and measure light.

In 2008, Politi et al. reported [94] the first demonstration of an *integrated quantum photonic chip*. The authors used established commercial fabrication techniques to construct complex linear-optical networks of beamsplitters on a cm-scale optical chip. These devices were shown to support high-fidelity [95] classical and quantum interference of single photons generated by SPDC, with a reported HOM-dip visibility of 1.001 ± 0.004 %. These were the first results in what is now a broad field of *integrated quantum photonics*. Other early demonstrations include on-chip quantum metrology [96] and a compiled implementation of Shor's factoring algorithm [74].

IQP provides a reduction in the scale of optical circuits, by at least an order of magnitude with respect to bulk optics. Moreover, monolithic integration provides operational advantages, one of the most significant of which is intrinsic stability of optical phase and mode-matching. This inherent stability has since allowed a number of demonstrations using path-encoded qubits, which in bulk optics are extremely susceptible to mechanical vibration and thermal drift. Moreover, owing to the degree of control and precision afforded by modern lithographic fabrication techniques, mode-matching at integrated beamsplitters can be very well-engineered, further improving the visibility of quantum and classical interference. These advantages in scale and stability immediately enabled the demonstration of quantum effects in circuits which would be unmanageably complex in a bulk-optical setup. Peruzzo [97] reported quantum walks of photon pairs in an array of 21 sites (see Sect. 6.3), as well as quantum interference in a 4-mode coupler [98]. Since the on-chip propagation distance can be much smaller than that of the equivalent bulk setup, integrated quantum photonic chips can also serve to reduce net photon loss, accelerating the speed at which experiments can be performed.

The first demonstrations of IQP used a lithographically-fabricated glass (silica)-based material system (Sect. 2.2.1). More recent demonstrations have highlighted the potential benefits of various alternative materials and fabrication techniques. Of particular interest is the prospect of integrated SPSs and single-photon detectors, together with classical digital electronics, potentially enabling a full quantum system-on-a-chip. Integrated spontaneous sources have been reported in silicon [56, 99] and lithium niobate [100]. Integration of optical waveguides with high-efficiency superconducting single-photon detectors was reported by Calkins et al. [101]. Increasingly sophisticated devices [102–104] have recently been fabricated using a direct-write technique [105], which also allows for three-dimensional waveguide structures [106, 107].

In the following section, we describe the design and implementation of a novel quantum photonic chip, incorporating two path-encoded qubits. We then go on to show the utility and flexibility of this chip in Chaps. 3–5. This device, if constructed in bulk, would occupy a full optical bench—clearly illustrating the significant practical advantage already afforded by IQP.

References

1. M.A. Nielsen, I.L. Chuang, in *Quantum Computation and Quantum Information (Cambridge Series on Information and the Natural Sciences)*, 1st edn. (Cambridge University Press, 2004)
2. J. Preskill, in *Quantum Information Lecture Notes* (Chapter 1)
3. P.A.M. Dirac, *The Principles of Quantum Mechanics*, 4th edn. (Oxford University Press, New York, 1982)
4. K. Hannabuss, in *An Introduction to Quantum Theory* (Oxford University Press, 1997)
5. S. Aaronson, in *Quantum Computing since Democritus* (Cambridge University Press, 2013)
6. A. Einstein, B. Podolsky, N. Rosen, Can quantum-mechanical description of physical reality be considered complete? Phys. Rev. Online Arch. (Prola) **47**, 777–780 (1935)
7. M.B. Plenio, S. Virmani, An introduction to entanglement measures. Quantum Inf. Comput. **7**, 1–51 (2006)
8. D.M. Greenberger, M.A. Horne, A. Shimony, A. Zeilinger, Bell's theorem without inequalities. Am. J. Phys. **58**(12) (1990)
9. C.H. Bennett, S. Popescu, B. Schumacher, Concentrating partial entanglement by local operations. Phys. Rev. A **53**(4), 2046–2052 (1996)
10. N. Brunner, D. Cavalcanti, S. Pironio, V. Scarani, S. Wehner, Bell nonlocality (2013). arXiv
11. J.S. Bell, On the Einstein Podolsky Rosen Paradox. Physics **1**(3), 195–200 (1964)
12. John F. Clauser, Michael A. Horne, Abner Shimony, Richard A. Holt, Proposed experiment to test local hidden-variable theories. Phys. Rev. Lett. **23**, 880–884 (1969)
13. S. Popescu, D. Rohrlich, Quantum nonlocality as an axiom. Found. Phys. **24**(3), 379–385 (1994)

14. B.S. Tsirelson, Some results and problems on quantum bell-type inequalities. Hadronic J. Suppl. **8**, 329–345 (1993)
15. Alain Aspect, Jean Dalibard, Gérard Roger, Experimental test of Bell's inequalities using time—varying analyzers. Phys. Rev. Lett. **49**(25), 1804–1807 (1982)
16. M. Giustina, A. Mech, S. Ramelow, B. Wittmann, J. Kofler, J. Beyer, A. Lita, B. Calkins, T. Gerrits, S.W. Nam, R. Ursin, A. Zeilinger, Bell violation using entangled photons without the fair-sampling assumption. Nature **497**, 227–230 (2013)
17. T. Scheidl, R. Ursin, J. Kofler, S. Ramelow, X.-S. Ma, T. Herbst, L. Ratschbacher, A. Fedrizzi, N.K. Langford, T. Jennewein, A. Zeilinger, Violation of local realism with freedom of choice. P. Natl. Acad. Sci. U.S.A. **107**(46), 19708–19713 (2010)
18. R.F. Werner, Quantum states with einstein-podolsky-rosen correlations admitting a hidden-variable model. Phys. Rev. A **40**, 4277–4281 (1989)
19. D. Deutsch, Quantum theory, the Church-turing principle and the universal quantum computer. Proc. Roy. Soc. Lond. A **400**, 97–117 (1985)
20. R.P. Feynman, Simulating physics with computers. Int. J. Theor. Phy. Theor. Phy. **21**, 467–488 (1982)
21. R.P. Feynman, Quantum mechanical computers. Found. Phys. **16**(6), 507–531 (1986)
22. P.W. Shor, Algorithms for quantum computation: discrete logarithms and factoring. In *SFCS '94: Proceedings of the* 35th *Annual Symposium on Foundations of Computer Science*, vol. 0 (IEEE Computer Society, Washington, 1994), pp. 124–134
23. D.P. DiVincenzo, Quantum computation. Science **270**(5234), 255–261 (1995)
24. A. Barenco, C.H. Bennett, R. Cleve, D.P. DiVincenzo, N. Margolus, P. Shor, T. Sleator, J.A. Smolin, H. Weinfurter, Elementary gates for quantum computation. Phys. Rev. A **52**, 3457–3467 (1995)
25. S. Lloyd, Almost any quantum logic gate is universal. Phys. Rev. Lett. **75**, 346–349 (1995)
26. P.O. Boykin, T. Mor, M. Pulver, V. Roychowdhury, F. Vatan, On universal and fault-tolerant quantum computing: a novel basis and a new constructive proof of universality for shor's basis. In *Foundations of Computer Science, 1999. 40th Annual Symposium on* (1999), pp. 486–494
27. D.P. DiVincenzo, The physical implementation of quantum computation. Fortschr. Phys. **48**, 771–783 (2000)
28. R. Raussendorf, H.J. Briegel, A one-way quantum computer. Phys. Rev. Lett. **86** (2001)
29. R. Raussendorf, D. Browne, H. Briegel, Measurement-based quantum computation on cluster states. Phys. Rev. A **68**(2) (2003)
30. P.W. Shor, Scheme for reducing decoherence in quantum memory. Phys. Rev. A **52**, 3457–3467 (1995)
31. S.J. Devitt, K. Nemoto, Programming a topological quantum computer (2012). ArXiv e-prints
32. D. Gottesman, An introduction to quantum error correction and Fault-Tolerant quantum computation (2009). ArXiv e-prints
33. C. Bennett, G. Brassard, Quantum cryptography: Public key distribution and coin tossing. Proc. IEEE Int. (1984)
34. L. Lydersen, C. Wiechers, C. Wittmann, D. Elser, J. Skaar, V. Makarov, Hacking commercial quantum cryptography systems by tailored bright illumination. Nat. Photonics **4**, 686–689 (2010)
35. S. Pironio, A. Acín, N. Brunner, N. Gisin, S. Massar, V. Scarani, Device-independent quantum key distribution secure against collective attacks. New J. Phys. **11**(4), 045021 (2009)
36. V. Giovannetti, S. Lloyd, L. Maccone, Quantum-enhanced measurements: beating the standard quantum limit. Science **306**, 1330–1336 (2004)
37. V. Giovannetti, S. Lloyd, L. Maccone, Advances in quantum metrology. Nat. Photonics **5**, 222–229 (2011)
38. J.C.F. Matthews, X.-Q. Zhou, H. Cable, P.J. Shadbolt, D.J. Saunders, G.A. Durkin, G.J. Pryde, J.L. O'Brien, Practical quantum metrology (2013). ArXiv e-prints
39. H. Cable, G.A. Durkin, Parameter estimation with entangled photons produced by parametric down-conversion. Phys. Rev. Lett. **105**, 013603 (2010)

40. P.S. Venkataram, Electromagnetic field quantization and applications to the casimir effect. MIT. Edu. (2013)
41. http://www.photond.com/products/fimmwave.htm
42. http://www.phoenixbv.com
43. R.J. Glauber, Coherent and incoherent states of the radiation field. Phys. Rev. **131**, 2766–2788 (1963)
44. R.H. Brown, R.Q. Twiss, A test of a new type of stellar interferometer on Sirius. Nature **178**, 1046–1048 (1956)
45. K.P. Zetie, S.F. Adams, R.M. Tocknell, How does a mach-zehnder interferometer work? Phys. Educ. **35**(1), 46 (2000)
46. P. Grangier, G. Roger, A. Aspect, Experimental evidence for a photon anticorrelation effect on a beam splitter: A new light on single-photon interferences. EPL (Europhys. Lett.) **1**(4), 173 (1986)
47. S.M. Tan, D.F. Walls, M.J. Collett, Nonlocality of a single photon. Phys. Rev. Lett. **66**, 252–255 (1991)
48. C.K. Hong, Z.Y. Ou, L. Mandel, Measurement of subpicosecond time intervals between two photons by interference. Phys. Rev. Lett. **59**, 2044–2046 (1987)
49. J. Rarity, P. Tapster, R. Loudon, Non-classical interference between independent sources (1997). arXiv:quant-ph/9702032
50. S. Scheel, Permanents in linear optical networks (2004). arXiv:quant-ph/0406127
51. E.R. Caianiello, On quantum field theory, 1: explicit solution of dyson's equation in electrodynamics without use of feynman graphs. Nuovo Cimento **10** (1953)
52. L.G. Valiant, The complexity of computing the permanent. Theor. Comput. Sci. **8**, 189–201 (1979)
53. T. Tilma, E.C.G. Sudarshan, Generalized Euler angle parametrization for SU(N). J. Phys. Math. Gen. **35**, 10467–10501 (2002)
54. E. Martín-López, A. Laing, T. Lawson, R. Alvarez, X.-Q. Zhou, J.L. O'Brien, Experimental realization of Shor's quantum factoring algorithm using qubit recycling. Nat. Photonics **6**, 773–776 (2012)
55. J.L. O'Brien, G.J. Pryde, A.G. White, T.C. Ralph, D. Branning, Demonstration of an all-optical quantum controlled-NOT gate. Nature **426**, 264–267 (2003)
56. J. Silverstone, D. Bonneau, K. Ohira, N. Suzuki, H. Yoshida, N. Iizuka, M. Ezaki, R. Hadfield, G.D. Marshall, V. Zwiller, J. Rarity, J. OBrien, M. Thompson, On-chip quantum interference between two silicon waveguide sources (2013). arXiv:1304.1490
57. M. Reck, A. Zeilinger, H.J. Bernstein, P. Bertani, Experimental realization of any discrete unitary operator. Phys. Rev. Lett. **73**, 58–61 (1994)
58. S. Aaronson, A. Arkhipov, The computational complexity of linear optics (2010). arXiv
59. N. Russell, E.M. López, A. Laing, In preparation
60. M.R. Geller, J.M. Martinis, A.T. Sornborger, P.C. Stancil, E.J. Pritchett, A. Galiautdinov, Universal quantum simulation with pre-threshold superconducting qubits: Single-excitation subspace method (2012). arXiv:1210.5260
61. R. Blatt, D. Wineland, Entangled states of trapped atomic ions. Nature **453**, 1008–1015 (2008)
62. B.P. Lanyon, P. Jurcevic, M. Zwerger, C. Hempel, E.A. Martinez, W. Dür, H.J. Briegel, R. Blatt, C.F. Roos, Measurement-based quantum computation with trapped ions. Phys. Rev. Lett. **111**, 210501 (2013)
63. M.H. Devoret, R.J. Schoelkopf, Superconducting circuits for quantum information: An outlook. Science **339**(6124), 1169–1174 (2013)
64. L. Robledo, L. Childress, H. Bernien, B. Hensen, P.F.A. Alkemade, R. Hanson, High-fidelity projective read-out of a solid-state spin quantum register. Nature **477**, 574–578 (2011)
65. B.E. Kane, A silicon-based nuclear spin quantum computer. Nature **393**(6681), 133–137 (1998)
66. T.F. Rønnow, Z. Wang, J. Job, S. Boixo, S.V. Isakov, D. Wecker, J.M. Martinis, D.A. Lidar, M. Troyer, Defining and detecting quantum speedup (2014). ArXiv e-prints

67. A.B. U'Ren, C. Silberhorn, K. Banaszek, I.A. Walmsley, Efficient conditional preparation of high-fidelity single photon states for fiber-optic quantum networks. Phys. Rev. Lett. **93**(9), 093601 (2004)
68. G.I. Taylor, Interference fringes with feeble light. Proc. Camb. Philos. Soc. **15**, 114–115 (1909)
69. X.-C. Yao, T.-X. Wang, X. Ping, L. He, G.-S. Pan, X.-H. Bao, C.-Z. Peng, L. Chao-Yang, Y.-A. Chen, J.-W. Pan, Observation of eight-photon entanglement. Nat. Photonics **6**(4), 225–228 (2012)
70. A.R. Dixon, Z.L. Yuan, J.F. Dynes, A.W. Sharpe, A.J. Shields, Gigahertz decoy quantum key distribution with 1 mbit/s secure key rate. Opt. Express **16**(23), 18790–18979 (2008)
71. J.G. Rarity, P.R. Tapster, E. Jakeman, T. Larchuk, R.A. Campos, M.C. Teich, B.E.A. Saleh, Two-photon interference in a mach-zehnder interferometer. Phys. Rev. Lett. **65**, 1348–1351 (1990)
72. L. Chao-Yang, D.E. Browne, T. Yang, J.-W. Pan, Demonstration of a compiled version of shor's quantum factoring algorithm using photonic qubits. Phys. Rev. Lett. **99**, 250504 (2007)
73. B.P. Lanyon, T.J. Weinhold, N.K. Langford, M. Barbieri, D.F.V. James, A. Gilchrist, A.G. White, Experimental demonstration of a compiled version of shor's algorithm with quantum entanglement. Phys. Rev. Lett. **99**, 250505+ (2007)
74. A. Politi, J.C.F. Matthews, J.L. O'Brien, Shor's quantum factoring algorithm on a photonic chip. Science **325**, 1221+ (2009)
75. S. Barz, R. Vasconcelos, C. Greganti, M. Zwerger, W. Dür, H.J. Briegel, P. Walther, Demonstrating an element of measurement-based quantum error correction (2013). ArXiv e-prints
76. X. Cai, J. Wang, M.J. Strain, B. Johnson-Morris, J. Zhu, M. Sorel, J.L. O'Brien, M.G. Thompson, Y. Siyuan, Integrated compact optical vortex beam emitters. Science **338**(6105), 363–366 (2012)
77. P. Kok, W.J. Munro, K. Nemoto, T.C. Ralph, J.P. Dowling, G.J. Milburn, Linear optical quantum computing (2005). arXiv, Review article
78. S.J. Devitt, A.D. Greentree, R. Ionicioiu, J.L. O'Brien, W.J. Munro, L.C.L. Hollenberg, Photonic module: An on-demand resource for photonic entanglement. Phys. Rev. A **76**(5), 052312 (2007)
79. E. Knill, R. Laflamme, G.J. Milburn, A scheme for efficient quantum computation with linear optics. Nature **409**, 46–52 (2001)
80. P. Kok, S.L. Braunstein, Limitations on the creation of maximal entanglement. Phys. Rev. A **62**(6), 064301 (2000)
81. J.L. O'Brien, A. Furusawa, J. Vuckovic, Photonic quantum technologies. Nat. Photon. **3**, 687–695 (2009)
82. S. Gasparoni, J.W. Pan, P. Walther, T. Rudolph, A. Zeilinger, Realization of a photonic controlled-NOT gate sufficient for quantum computation. Phys. Rev. Lett. **93**, 020504+ (2004)
83. M.A. Nielsen, Optical quantum computation using cluster states. arXiv (4) (2004)
84. D.E. Browne, T. Rudolph, Resource-efficient linear optical quantum computation. Phys. Rev. Lett. **95**(1) (2005)
85. P. Walther, K.J. Resch, T. Rudolph, E. Schenck, H. Weinfurter, V. Vedral, M. Aspelmeyer, A. Zeilinger, Experimental one-way quantum computing. arXiv (2005)
86. R. Prevedel, P. Walther, F. Tiefenbacher, P. Bohi, R. Kaltenbaek, T. Jennewein, A. Zeilinger, High-speed linear optics quantum computing using active feed-forward. Nature **445**, 65–69 (2007)
87. R. Ceccarelli, G. Vallone, F. De Martini, P. Mataloni, A. Cabello, Experimental entanglement and nonlocality of a Two-Photon Six-Qubit cluster state. Phys. Rev. Lett. **103**, 160401+ (2009)
88. T. Horikiri, H. Sasaki, H. Wang, T. Kobayashi, Security and gain improvement of a practical quantum key distribution using a gated single-photon source and probabilistic photon-number resolution. Phys. Rev. A **72**, 012312 (2005)
89. M.D. Eisaman, J. Fan, A. Migdall, S.V. Polyakov, Invited review article: Single-photon sources and detectors. Rev. Sci. Instrum. **82**(7) (2011)

90. M.J. Collins, C. Xiong, I.H. Rey, T.D. Vo, J. He, S. Shahnia, C. Reardon, T.F. Krauss, M.J. Steel, A.S. Clark, B.J. Eggleton, Integrated spatial multiplexing of heralded single-photon sources. Nat. Commun. **4** (2013)
91. E. Jeffrey, N.A. Peters, P.G. Kwiat, Towards a periodic deterministic source of arbitrary single-photon states. New J. Phys. **6**(1), 100 (2004)
92. L. Mandel, E. Wolf, in *Optical coherence and quantum Optics* (Cambridge University Press, 1995)
93. P.G. Kwiat, K. Mattle, H. Weinfurter, A. Zeilinger, A.V. Sergienko, Y. Shih, New high-intensity source of polarization-entangled photon pairs. Phys. Rev. Lett. **75**, 4337–4341 (1995)
94. A. Politi, M.J. Cryan, J.G. Rarity, Y. Siyuan, J.L. O'Brien, Silica-on-Silicon waveguide quantum circuits. Science **320**, 646–649 (2008)
95. A. Laing, A. Peruzzo, A. Politi, M.R. Verde, M. Halder, T.C. Ralph, M.G. Thompson, J.L. O'Brien, High-fidelity operation of quantum photonic circuits. Appl. Phys. Lett. **97**, 211109+ (2010)
96. J.C.F. Matthews, A. Politi, A. Stefanov, J.L. O'Brien, Manipulation of multiphoton entanglement in waveguide quantum circuits. Nat. Photon. **3**, 346–350 (2009)
97. A. Peruzzo, M. Lobino, J.C.F. Matthews, N. Matsuda, A. Politi, K. Poulios, X.-Q. Zhou, Y. Lahini, N. Ismail, K. Wörhoff, Y. Bromberg, Y. Silberberg, M.G. Thompson, J.L. OBrien, Quantum walks of correlated photons. Science **329**, 1500–1503 (2010)
98. A. Peruzzo, A. Laing, A. Politi, T. Rudolph, J.L. O'Brien, Multimode quantum interference of photons in multiport integrated devices. Nat. Commun. **2**, 224+ (2011)
99. N. Matsuda, H. Le Jeannic, H. Fukuda, T. Tsuchizawa, W.J. Munro, K. Shimizu, K. Yamada, Y. Tokura, H. Takesue, A monolithically integrated polarization entangled photon pair source on a silicon chip. Sci. Rep. **2** (2012)
100. W. Sohler, H. Hui, R. Ricken, V. Quiring, C. Vannahme, H. Herrmann, D. Büchter, S. Reza, W. Grundkötter, S. Orlov, H. Suche, R. Nouroozi, Y. Min, Integrated optical devices in lithium niobate. Opt. Photon. News **19**(1), 24–31 (2008)
101. B. Calkins, P.L. Mennea, A.E. Lita, B.J. Metcalf, W.S. Kolthammer, A. Lamas-Linares, J.B. Spring, P.C. Humphreys, R.P. Mirin, J.C. Gates, P.G.R. Smith, I.A. Walmsley, T. Gerrits, S.W. Nam, High quantum-efficiency photon-number-resolving detector for photonic on-chip information processing. Opt. Express **21**, 22657 (2013)
102. N. Spagnolo, C. Vitelli, L. Sansoni, E. Maiorino, P. Mataloni, F. Sciarrino, D.J. Brod, E.F. Galvao, A. Crespi, R. Ramponi, R. Osellame, General rules for bosonic bunching in multimode interferometers. Phys. Rev. Lett. **111**(13), 130503 (2013)
103. A. Crespi, R. Ramponi, R. Osellame, L. Sansoni, I. Bongioanni, F. Sciarrino, G. Vallone, P. Mataloni, Integrated photonic quantum gates for polarization qubits. Nat. Commun. **2**, 566 (2011)
104. M. Tillmann, B. Dakić, R. Heilmann, S. Nolte, A. Szameit, P. Walther, Experimental Boson sampling. Nat. Photonics **7**(7), 540–544 (2013)
105. D.G. Marshall, A. Politi, J.C.F. Matthews, P. Dekker, M. Ams, M.J. Withford, J.L. O'Brien, Laser written waveguide photonic quantum circuits. Opt. Express **17**, 12546–12554 (2009)
106. K. Poulios, R. Keil, D. Fry, J.D.A. Meinecke, J.C.F. Matthews, A. Politi, M. Lobino, M. Gräfe, M. Heinrich, S. Nolte, A. Szameit, J.L. O'Brien, Quantum walks of correlated photon pairs in two-dimensional waveguide arrays (2013). arXiv:1308.2554
107. M.C. Rechtsman, J.M. Zeuner, A. Tünnermann, S. Nolte, M. Segev, A. Szameit, Strain-induced pseudomagnetic field and photonic Landau levels in dielectric structures. Nat. Photonics **7**, 153–158 (2013)

Chapter 2
A Reconfigurable Two-Qubit Chip

Machines take me by surprise with great frequency.

Alan Turing

2.1 Introduction

The discovery and development of universal computing machines is one of the greatest scientific accomplishments of the 20th century. The Church-Turing thesis—that all calculable functions can be computed by a particularly simple type of machine—is generally expressed as a statement about mathematical functions, and the evaluation of numbers. However, the influence of universal computing machines has stretched much further than the academic mathematical context in which they were first conceived, having profound effects on social, economic and artistic life.

The prospective benefits of quantum computing enjoy a similar promise of universality. Specifically, we believe [1] that a scalable machine satisfying the DiVincenzo criteria (Sect. 1.4.1) would be universal for quantum computing, and could run any quantum algorithm, prepare any quantum state or operator,[1] and would also be universal for classical computation. This promise allows us to progress with the development of the basic building blocks of quantum information technologies, without complete information on the potential applications of quantum computing: although we have a small number of specific examples of quantum algorithms which provide an exponential speedup over classical machines, it is reasonable to think that, as with classical computation, the scope of useful applications will ultimately prove to be much broader.

The results of KLM (Sect. 1.6.2), together with more recent developments in cluster-state theories [2–5], show that in principle LOQC can provide a scalable route

[1]Note that this does not imply any particular scaling: arbitrary N-qubit state preparation is exponentially hard even for quantum computers. See Chap. 5 for further discussion.

P. Shadbolt, *Complexity and Control in Quantum Photonics*,
Springer Theses, DOI 10.1007/978-3-319-21518-1_2

to universal quantum computation. More recently, integrated quantum photonics (Sect. 1.6.5) has been shown to offer an *experimentally* scalable approach to the construction of LOQC machines, potentially allowing millions [6] of components to be lithographically fabricated on a single monolithic chip. Early results in the field include the demonstration of quantum interference in passive linear optical interferometers [7–10], as well as active devices with reconfigurable phase shifters [11–13]. Notably, most of these reconfigurable devices used a single phase shifter, giving the device a single classical control parameter. This was sufficient for novel demonstrations of quantum metrology [11] and switching of entangled photonic states [13]. However, much of the utility and interest of a universal quantum computer arises from the fact that a single machine can be arbitrarily reconfigured to perform a broad variety of tasks. This degree of reconfigurability requires a large (polynomial) number of classical control parameters, and is the main focus of work described in this section.

We describe a waveguide linear-optical circuit which can encode and manipulate the state of two photonic qubits using two indistinguishable photons from an SPDC source. This device features eight voltage-controlled phase shifters, which can be arbitrarily reconfigured to prepare any two-qubit state. The architecture of the device includes four reconfigurable single-qubit operations, together with a passive two-qubit entangling gate. As such, the gate operations implemented in this device comprise a universal quantum gate set (Sect. 1.4.1).

In close analogy with classical computers, we find that the degree of reconfigurability afforded by this device has allowed a surprisingly rich variety of physical phenomena and quantum information techniques to be studied, above and beyond the original intent of the device. Indeed, Chaps. 3, 4, and 5 all make use of the two-qubit chip described here. This work highlights the fact that nontrivial experiments can be performed using even a very small number of qubits, in contrast with the classical case—where the scope of worthwhile experiments using only two classical bits is limited.

To our knowledge, this work includes the first experimental implementation of photonic two-qubit quantum state and process tomography (where state preparation and measurement were performed on-chip), and the first photonic on-chip Bell inequality violation.

2.2 CNOT-MZ

The CNOT-MZ is a reconfigurable quantum photonic chip, shown schematically in Fig. 2.1b. Two qubits are encoded in path, using indistinguishable photon pairs at 808 nm, generated by type-I SPDC. The chip uses a total of 6 waveguides, 13 directional couplers and 8 voltage-controlled thermal phaseshifters to implement the circuit model diagram shown in Fig. 2.1a.

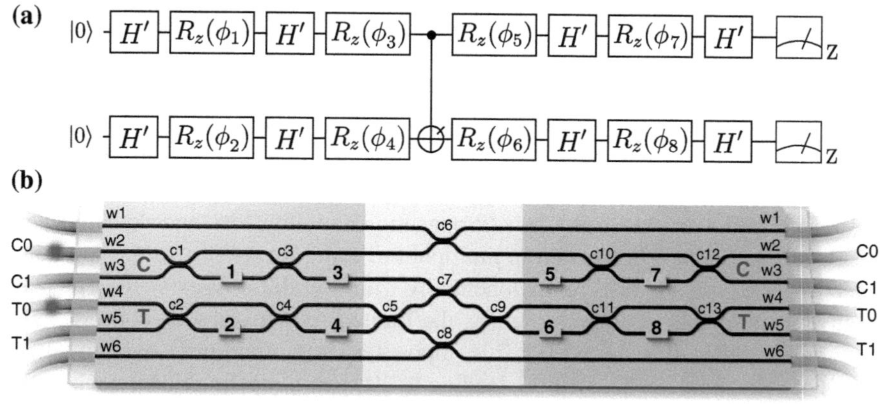

Fig. 2.1 CNOT-MZ chip. **a** Circuit-model diagram. Two qubits are prepared in the $|00\rangle$ state. H' is a Hadamard-like gate corresponding to a directional coupler, with the same unitary matrix representation as a beamsplitter (1.119). $R_z(\phi)$ correspond to voltage-controlled phase shifts, and implement single-qubit rotations about the z-axis of the Bloch sphere (1.148). At the centre of the chip is a two-qubit CNOT-P entangling gate, locally equivalent to the maximally entangling CNOT gate. Each qubit can be effectively measured in an arbitrary basis, by combining single-photon rotations with measurement in the z-basis. **b** Waveguide architecture. All DCs have coupling ratio $\eta = 1/2$, apart from c_6, c_7 and c_8, which are engineered to transmit a fraction $\eta = t = 2/3$ of incident light. Two indistinguishable photons generated by type-I SPDC are coupled into the chip, and encode two qubits in path. Waveguides $w_{2,3}$ and $w_{4,5}$ correspond to the $|0\rangle$ and $|1\rangle$ states of the control and target qubit respectively. Waveguides w_1 and w_6 do not correspond to logical basis states. The first stage of the chip uses two MZIs and four phaseshifters to implement arbitrary two-qubit separable state preparation. The central section implements the CNOT-P gate. The final section of the chip uses two MZIs, together with off-chip single-photon detection, to implement arbitrary separable two-qubit measurements

The architecture of the chip is based around a passive postselected linear-optical CNOT(CNOT-P) gate, which implements a maximally entangling CNOT-like operation on the two qubits. This gate is discussed in detail in Sect. 2.2.4. The control qubit is encoded using waveguides w_2 and w_3, corresponding to the $|0\rangle$ and $|1\rangle$ states respectively, and the target qubit is similarly encoded across w_4 and w_5. Each qubit is initially prepared in the $|0\rangle$ state, with photon pairs coupled directly from the source into waveguides w_2 and w_4. Arbitrary state preparation of each qubit is then accomplished using an MZI with two phaseshifters, as described in Sect. 2.2.5. At the output of the CNOT-P gate, each qubit is measured in a local basis using an MZI together with two single-photon detectors, as described in Sect. 2.2.6.

The device was fabricated by CIP technologies [14] in a silica-based material system, described in Sect. 2.2.1. The chip die is 3 mm wide and 70 mm long. Full details of the photon source, control system and supporting experimental setup are given in Sect. 2.3.

2.2.1 Silica-on-Silicon

Glass (silica) waveguides are particularly well-suited for quantum applications. In particular, they exhibit very low propagation loss (<0.1 dB cm^{-1}), couple well to single-mode optical fibre (typically ~70 % coupling efficiency), and are transparent to the band around ~800 nm where SPDC sources and room-temperature APD single-photon detectors are most efficient. Propagation/coupling loss and detection efficiency are particularly important in multiphoton experiments, where the N-photon detection rate typically falls off exponentially with overall loss η as $1/\eta^N$. The main disadvantage of this material system is the limited refractive index contrast, typically on the order of $\Delta = 0.5\,\%$. This imposes a large minimum waveguide bend radius of ~15 mm (see Sect. 1.5.1), leading to 200 µm-wide directional couplers on the order of ~6 mm in length. Recently, more compact devices have been achieved using alternative material systems, at the cost of greater loss (see Sects. 6.3.3, 6.3.4 and 2.10).

The CNOT-MZ device was fabricated using silica-on-silicon planar lightwave circuit technology, shown in Fig. 2.2. A 16 µm buffer layer of undoped silica was grown on a silicon substrate, forming the lower cladding of the waveguides. A 3.5 µm layer of silica doped with germanium and boron oxides was overgrown, and was then lithographically etched to form the square 3.5 µm × 3.5 µm waveguide core, with a refractive index contrast between core and cladding of $\Delta = 0.5\,\%$. A 16 µm-thick upper cladding of silica, doped with phosphorous and boron to match the lower cladding, was then overgrown. Finally, a metallic layer was deposited and lithographically etched to form resistive heaters, electrical connections, and probe contact pads on the top surface of the chip.

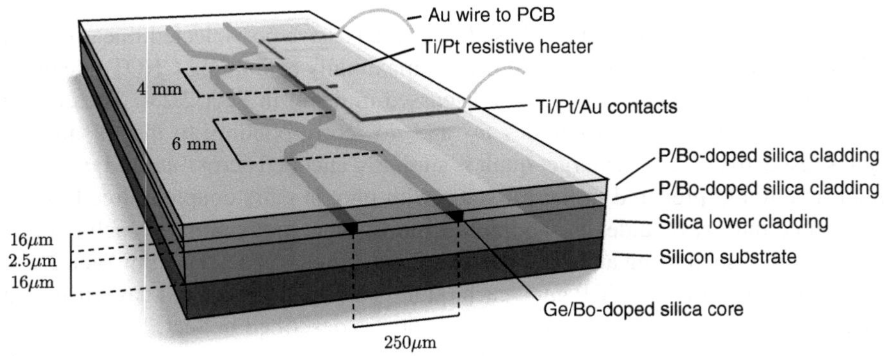

Fig. 2.2 Silica-on-silicon material system and waveguide geometry. Square 2.5 µm × 2.5 µm waveguides were fabricated in germanium/boron-doped silica, on a silicon substrate. The waveguide cladding is a combination of undoped silica and phosphorous/boron-doped silica. Titanium/platinum/gold traces connect to titanium/platinum resistive heaters, allowing a reconfigurable voltage-controlled phase shift to be applied. Contact pads were gold-wire-bonded to a standard PCB

The waveguides used here have a symmetric (square) profile, which together with the amorphous, isotropic nature of silica leads to negligible birefringence. As a result, in principle these waveguides will support any single polarization of light. Although on-chip polarization encoding has been demonstrated in a number of material systems [13, 15, 16], it remains challenging—in particular due to unwanted rotations introduced by waveguide bends—and in this work we operate in vertical polarization only.

2.2.2 Directional Coupler

Leading approaches to the implementation of two-mode beamsplitter operations in integrated photonics include multimode interference (MMI) couplers and DCs. Here we consider the latter, illustrated in Fig. 2.3, in which two waveguides are brought close together so as to couple the guided modes via the evanescent field (Sect. 1.5.1). Any DC is characterised by its coupling ratio η, corresponding to the fraction of optical power transmitted from one waveguide to the other, which is equivalent to the BS transmissivity (Sect. 1.5.2) and is controlled by the separation distance s and length L of the coupling region.

Mode coupling theory [17] gives the relationship between the field amplitude (1.66) in two coupled waveguides A, B as a system of coupled differential equations

$$\frac{dA(z)}{dz} = -i\kappa B(z); \quad \frac{dB(z)}{dz} = -i\kappa A(z), \tag{2.1}$$

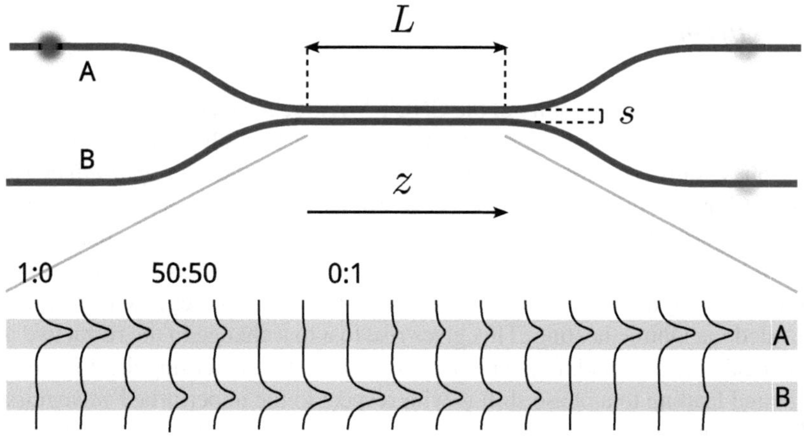

Fig. 2.3 Geometry of a directional coupler. Two waveguides are adiabatically brought into close proximity, such that the evanescent fields overlap. Light periodically couples from one waveguide to the other as a function of the propagation distance L and the coupling constant κ, which depends in part on the spatial separation s and refractive index n

where κ is a coupling constant which depends on the spatial overlap of the two guided modes. This leads to solutions of the form

$$A(z) = A_0 \cos(\kappa z) - B_0 i \sin(\kappa z); \quad B(z) = B_0 \cos(\kappa z) - A_0 i \sin(\kappa z), \qquad (2.2)$$

where A_0, B_0 are the initial field amplitudes at the input ports. As a result, in the coupling region of the DC, optical power oscillates sinusoidally between the two waveguides as a function of the interaction length L. By tuning this length, the DC can be designed to implement an arbitrary BS operation (1.119)

$$\begin{bmatrix} A(L) \\ B(L) \end{bmatrix} = \begin{bmatrix} \cos(\kappa L) & -i \sin(\kappa L) \\ -i \sin(\kappa L) & \cos(\kappa L) \end{bmatrix} \begin{bmatrix} A_0 \\ B_0 \end{bmatrix} = \Lambda_{\mathrm{DC}}(\kappa, L) \begin{bmatrix} A_0 \\ B_0 \end{bmatrix} = \begin{bmatrix} \sqrt{t} & i\sqrt{r} \\ i\sqrt{r} & \sqrt{t} \end{bmatrix}. \qquad (2.3)$$

In order to obtain a 50:50 DC with $t = r = \frac{1}{2}$, we must therefore have $L = \pi/4\kappa$, which for the silica-on-silicon material system used here, with $s = 3\,\mu\mathrm{m}$, corresponds to an interaction length of ~4 mm.

The quality of fabrication of directional couplers is critical to the performance of the reconfigurable two-qubit chip (CONTZ-MZ) and other linear-optical quantum circuits described in this thesis. Deviation from the designed coupling ratio leads to unitary errors in qubit state preparation and measurement, and reduces the contrast of classical interference. Moreover, errors in both coupling ratio and imperfect mode-matching at the interaction region of the coupler lead to reduced visibility of HOM interference, and thus contribute to the observed sub-unit quantum state/process fidelities reported in Sects. 2.6 and 2.7 of this thesis.

2.2.3 Thermal Phaseshifter

The general-purpose flexibility of the CNOT-MZ is achieved through the inclusion of eight reconfigurable phase shifters, as shown in Fig. 2.1. In silica-on-silicon, reconfigurable phase shifts are most easily implemented using the *thermo-optic effect*. Here, a metallic (titanium/platinum) resistive heater of length L is lithographically patterned on the top surface of the upper waveguide cladding, directly above the waveguide core, as shown in Fig. 2.2. This heater is connected via Ti/Pt/Au electrodes to a current source, allowing the temperature of a local region of the waveguide to be precisely controlled via Ohmic heating. This gives rise to a to a change in the refractive index of the local core and cladding, with $dn/dT \sim 10^{-5}$/K, increasing the effective path length and leading to a phase shift φ with respect to the unperturbed waveguide.

The maximum temperature difference supported by the silica-on-silicon material system is ~30 °C, and in order to achieve a range in phase of 2π the resistive heater must therefore have a length on the order of 4 mm. The heaters are rated for a maximum voltage of 5 V, however in order to achieve a full 2π phaseshift in all MZIs we had to exceed this limit, running most phaseshifters between 0 and 7 V, leading to a total of ~1 W per heater at maximum voltage. I-V curves for each resistive heater on the CNOT-MZ are shown in Fig. 2.8d, showing a typical resistance of $R \sim 60\,\Omega$. Further details of phaseshifter calibration are given in Sect. 2.3.3.

The main drawback of thermal phase shifting is switching speed: in the silica-on-silicon platform used here, heating/cooling of a phaseshifter for a differential phaseshift of π takes at least ~100 ms (Fig. 2.8c, inset). This limits the scope of applications—for instance, active feed-forward is not possible using this technology. However, in the majority of experiments described in this thesis, the time taken to acquire a sufficient number of single-photon detection events, corresponding to a single measurement of an expectation value, is typically at least 1 s, and in practice there was not any need to switch phases faster than 1 Hz. Alternative material systems for integrated quantum photonics support an electro-optic effect, where phase can switched electrically up to GHz frequencies. See [13] for an example in lithium niobate.

2.2.4 Linear-Optical CNOT-P Gate

It was shown by Lloyd [18] that almost any two-qubit entangling gate is universal for quantum computing, and by DiVincenzo that a universal gate set can always be constructed from a two-qubit entangling gate together with single-qubit rotations [19]. We have seen in Sect. 1.5.4 that deterministic, arbitrary single-qubit rotations are very easily constructed using linear optics. However, since photons do not interact, the greatest challenge (and the greatest accomplishment of KLM), is to find a scalable two-qubit entangling gate. All scalable approaches to linear optical quantum computing (LOQC), including KLM and more recent cluster-state techniques (Sects. 1.6.2 and 1.4.1), depend on active feed-forward. At the time of writing, although fast switching, low propagation loss, high refractive-index contrast, integrated GHz logic and single-photon detectors, etc. have all been demonstrated in separate photonic devices, no existing technology or material system satisfies all necessary conditions for a full demonstration of scalable LOQC with active feed-forward. Certainly, the thermal phase-shifters previously described are too slow for such applications.

In 2002, two groups [20, 21] proposed a scheme by which a two-qubit maximally-entangling gate can be implemented using linear-optics and postselection, without any need for feed-forward. It has already been stated (Sect. 1.6.2) that LOQS is not scalable without feed-forward, and indeed this gate does not scale—successful operation of the gate is postselected with probability 1/9, leading to exponentially decreasing success probability for composite circuits. However, the scheme is experimentally much more accessible, and an experimental demonstration was almost immediately reported by a number of groups [22–25]. An important property of the design of this postselected gate is that it possesses many of the same experimental prerequisites—indistinguishable photons, high visibility classical and quantum interference, stable interferometers—as the scalable CZ gate of KLM, and experimental implementations of the former thus constitute real progress towards the latter.

We will now sketch the basic mechanism of the postselected two-qubit gate, starting from an implementation of the CZ gate. CZ is a maximally entangling gate, which flips the sign of the target qubit when both input qubits are in the state $|1\rangle$:

$$\begin{aligned} |0_C 0_T\rangle_{\text{in}} &\rightarrow |0_C 0_T\rangle_{\text{out}}, \quad |0_C 1_T\rangle_{\text{in}} \rightarrow |0_C 1_T\rangle_{\text{out}}, \\ |1_C 0_T\rangle_{\text{in}} &\rightarrow |1_C 0_T\rangle_{\text{out}}, \quad |1_C 1_T\rangle_{\text{in}} \rightarrow -|1_C 1_T\rangle_{\text{out}}, \end{aligned} \tag{2.4}$$

and is therefore described by a unitary operator

$$\hat{U}_{\text{CZ}} = \begin{bmatrix} 1 & 0 & 0 & 0 \\ 0 & 1 & 0 & 0 \\ 0 & 0 & 1 & 0 \\ 0 & 0 & 0 & -1 \end{bmatrix}. \tag{2.5}$$

In order to see how this gate can be implemented in linear optics, it will be instructive to first consider the circuit shown in Fig. 2.4a. If two photons are injected into modes C_0 and T_0, encoding the logical input state $|0_C 0_T\rangle_{\text{in}}$, the resulting evolution is trivial

$$|0_C 0_T\rangle_{\text{in}} = \hat{a}^\dagger_{\text{C}_0} \hat{a}^\dagger_{\text{T}_0} |\mathbf{0}\rangle \rightarrow (i\hat{a}^\dagger_{\text{C}'_0})(i\hat{a}^\dagger_{\text{T}'_0})|\mathbf{0}\rangle = -|1_{\text{C}'_0} 0_{\text{C}'_1} 1_{\text{T}'_0} 1_{\text{T}'_1}\rangle = -|0_C 0_T\rangle_{\text{out}}, \tag{2.6}$$

where the phase i arises from reflection at the mirrors. Similarly, for input states $|0_C 1_T\rangle_{\text{in}}$ and $|1_C 0_T\rangle_{\text{in}}$ the two photons never meet, and the system evolves as

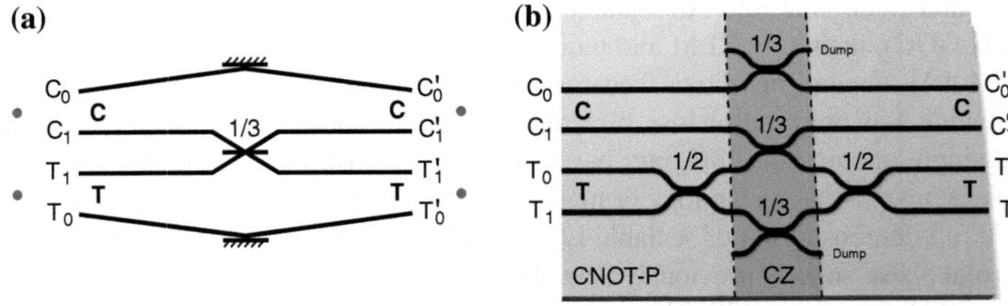

Fig. 2.4 CNOT-P gate construction. **a** Postselected linear-optical CZ gate, without dump modes. Control and target qubits are encoded in path, using indistinguishable single photon pairs. Postselecting on detection events in the two-qubit subspace, quantum interference at the 1/3-reflectivity beamsplitter gives rise to a relative phase shift of -1 for the $|11\rangle$ input state. Note that as shown, the effective gate operation after postselection is not unitary. **b** Waveguide implementation of a linear-optical CNOP-T gate. 1/3-reflectivity DCs in the central region of the device implement a CZ gate, where the top and bottom couplers "dump" probability amplitude, avoiding the non-unitarity of the device shown in (**a**). By adding two 1/2-reflectivity DCs to the target qubit, the CZ gate is converted to a CNOT-like gate, acting on the logical basis. This gate forms the basis for the CNOT-MZ circuit, Fig. 2.1

$$
\begin{aligned}
|0_C 1_T\rangle_{\text{in}} = \hat{a}^\dagger_{\text{C}_0}\hat{a}^\dagger_{\text{T}_1}|\mathbf{0}\rangle &\rightarrow (i\hat{a}^\dagger_{\text{C}'_0})(i\sqrt{r}\,\hat{a}^\dagger_{\text{T}'_1} + \sqrt{t}\,\hat{a}^\dagger_{\text{C}'_1})|\mathbf{0}\rangle \\
&= -\left(\sqrt{r}\,|1_{\text{C}'_0}0_{\text{C}'_1}0_{\text{T}'_0}1_{\text{T}'_1}\rangle + \sqrt{t}\,|1_{\text{C}'_0}1_{\text{C}'_1}0_{\text{T}'_0}0_{\text{T}'_1}\rangle\right), \quad (2.7) \\
|1_C 0_T\rangle_{\text{in}} = \hat{a}^\dagger_{\text{C}_1}\hat{a}^\dagger_{\text{T}_0}|\mathbf{0}\rangle &\rightarrow (i\sqrt{r}\,\hat{a}^\dagger_{\text{C}'_1} + \sqrt{t}\,\hat{a}^\dagger_{\text{T}'_1})(i\hat{a}^\dagger_{\text{T}'_0})|\mathbf{0}\rangle \\
&= -\left(\sqrt{r}\,|0_{\text{C}'_0}1_{\text{C}'_1}1_{\text{T}'_0}0_{\text{T}'_1}\rangle + \sqrt{t}\,|0_{\text{C}'_0}0_{\text{C}'_1}1_{\text{T}'_0}1_{\text{T}'_1}\rangle\right), \quad (2.8)
\end{aligned}
$$

where further phases i arise from reflection at the BS. Now, the Fock states $|1_{\text{C}'_0}1_{\text{C}'_1}0_{\text{T}'_0}0_{\text{T}'_1}\rangle$ and $|0_{\text{C}'_0}0_{\text{C}'_1}1_{\text{T}'_0}1_{\text{T}'_1}\rangle$ have both photons occupying the same qubit, and do not have a representation in the two-qubit encoding. We must therefore post-select on the two-qubit subspace, resulting in the effective evolution

$$
|0_C 1_T\rangle_{\text{in}} \rightarrow -\sqrt{r}\,|0_C 1_T\rangle_{\text{out}}; \quad |1_C 0_T\rangle_{\text{in}} \rightarrow -\sqrt{r}\,|1_C 0_T\rangle_{\text{out}}. \tag{2.9}
$$

When the input state is $|1_C 1_T\rangle_{\text{in}}$, the two photons meet at the beamsplitter and undergo quantum interference as described in Sect. 1.5.3. The system then evolves as

$$
|1_C 1_T\rangle_{\text{in}} = \hat{a}^\dagger_{\text{C}_1}\hat{a}^\dagger_{\text{T}_1}|\mathbf{0}\rangle \rightarrow \left(i\sqrt{r}\,\hat{a}^\dagger_{\text{C}_1} + \sqrt{t}\,\hat{a}^\dagger_{\text{T}_1}\right)\left(i\sqrt{r}\,\hat{a}^\dagger_{\text{T}_1} + \sqrt{t}\,\hat{a}^\dagger_{\text{C}_1}\right)|\mathbf{0}\rangle, \tag{2.10}
$$

$$
|\psi\rangle_{\text{out}} = \left((t-r)\hat{a}^\dagger_{\text{C}_1}\hat{a}^\dagger_{\text{T}_1} + i\sqrt{r}\sqrt{t}\,\hat{a}^\dagger_{\text{C}_1}\hat{a}^\dagger_{\text{C}_1} + i\sqrt{t}\sqrt{r}\,\hat{a}^\dagger_{\text{T}_1}\hat{a}^\dagger_{\text{T}_1}\right)|\mathbf{0}\rangle, \tag{2.11}
$$

where we have used the relation $\left[\hat{a}^\dagger_{\text{C}_1}, \hat{a}^\dagger_{\text{T}_1}\right] = 0$, since the two photons are indistinguishable. Postselecting on the C_1T_1 term, which is the only component corresponding to a two-qubit state, we find

$$
|1_C 1_T\rangle_{\text{in}} \rightarrow (t-r)|1_C 1_T\rangle_{\text{out}} \tag{2.12}
$$

Setting $r = 1 - t = 1/3$, we arrive at

$$
\begin{aligned}
&|0_C 0_T\rangle_{\text{in}} \rightarrow -|0_C 0_T\rangle_{\text{out}}, \quad |0_C 1_T\rangle_{\text{in}} \rightarrow \frac{-1}{\sqrt{3}}|0_C 1_T\rangle_{\text{out}}, \\
&|1_C 0_T\rangle_{\text{in}} \rightarrow \frac{-1}{\sqrt{3}}|1_C 0_T\rangle_{\text{out}}, \quad |1_C 1_T\rangle_{\text{in}} \rightarrow \frac{1}{3}|1_C 1_T\rangle_{\text{out}}.
\end{aligned} \tag{2.13}
$$

Neglecting the global phase of -1, we have then accomplished the essential function of the CZ gate: a conditional phaseshift by -1 of the $|1_C 1_T\rangle$ term only. However, this postselected operation does not correspond to a unitary operator on the qubit subspace, and is clearly biased towards the $|0_C 0_T\rangle$ state. To overcome this issue, we simply replace the mirrors shown in Fig. 2.4a with $1/3$-reflectivity beamsplitters. It is easy to see that this has the effect of multiplying the amplitudes of the $|0_C 0_T\rangle$, $|0_C 1_T\rangle$ and $|1_C 0_T\rangle$ terms by factors of $1/3$, $1/\sqrt{3}$ and $1/\sqrt{3}$ respectively, balancing the gate, and restoring unitarity. The circuit then exactly reproduces the behaviour of the CZ gate, conditional on detection of one photon in C_0 or C_1 and one photon T_0

or T_1. By the Born rule, this occurs with probability $1/9$. It has been shown that this success probability is optimal for linear-optical two-qubit gates of this type [26]. A waveguide implementation is shown in the center of Fig. 2.4b.

The CZ gate together with local rotations is universal for quantum computing. However, the CNOT gate, which is the quantum equivalent of a classical reversible reversible exclusive-OR (XOR) gate, is often conceptually easier to handle than CZ. The CNOT gate flips the state of the target qubit, conditional on the state of the control

$$|0_C 0_T\rangle \to |0_C 0_T\rangle, \quad |0_C 1_T\rangle \to |0_C 1_T\rangle, \quad |1_C 0_T\rangle \to |1_C 1_T\rangle, \quad |1_C 1_T\rangle \to |1_C 0_T\rangle. \tag{2.14}$$

Starting from the CZ gate, this is easily constructed by the addition of two single-qubit Hadamard operations

$$\hat{U}_{\text{CNOT}} = \begin{bmatrix} 1\,0\,0\,0 \\ 0\,1\,0\,0 \\ 0\,0\,0\,1 \\ 0\,0\,1\,0 \end{bmatrix} = \left(\mathbf{1} \otimes \hat{H}\right) \hat{U}_{\text{CZ}} \left(\mathbf{1} \otimes \hat{H}\right). \tag{2.15}$$

Since the single-qubit Hadamard gate is almost equivalent to a beamsplitter operation, this leads to a natural construction of the linear-optical CNOT gate by the addition of two $1/2$-reflectivity beamsplitters or DCs, as shown in Fig. 2.4b. Note that this gate does not exactly reproduce the two-qubit unitary $\hat{U}_{\text{CNOT}}$, instead implementing the locally equivalent operation

$$\hat{U}_{\text{CNOT}-\text{P}} = \left(\mathbf{1} \otimes \hat{U}_{\text{BS}}\right) \hat{U}_{\text{CZ}} \left(\mathbf{1} \otimes \hat{U}_{\text{BS}}\right) = \begin{bmatrix} 0 & i & 0 & 0 \\ i & 0 & 0 & 0 \\ 0 & 0 & 1 & 0 \\ 0 & 0 & 0 & -1 \end{bmatrix}. \tag{2.16}$$

As such we will refer to this postselected gate operation generated by the circuit in Fig. 2.4b as CNOT-P, to distinguish from the canonical CNOT gate. This gate was demonstrated in bulk optics by a number of groups [22–24, 27]. More recently, the CNOT-P was implemented in a silica-on-silicon integrated platform [7], and formed the basis for a linear-optical implementation of Shor's factoring algorithm [28].

It is important to emphasize that the basic mechanism of the CNOT-P gate depends necessarily on two-photon quantum interference, and that the gate fails if the input photon pair is made distinguishable.

2.2.5 State Preparation

The first stage of the CNOT-MZ is used to prepare two qubits in an arbitrary separable state. Two photons from the source are always injected into waveguides i_2 and i_4

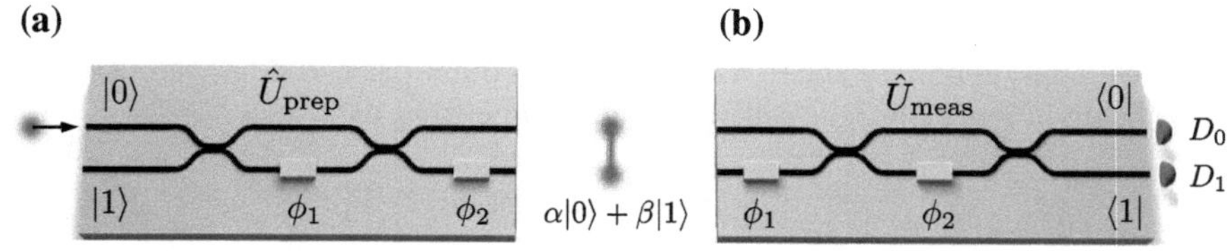

Fig. 2.5 State preparation and measurement of a single path-encoded qubit in linear optics

respectively, encoding the state $|00\rangle$. Each qubit is then acted upon by an MZI with two phaseshifters ϕ_1, ϕ_2 (Fig. 2.5a). We have already seen that an MZI with three phaseshifters is adequate for arbitrary single-qubit $SU(2)$ rotations. With the $|0\rangle$ state as input, two phaseshifters are sufficient for arbitrary state preparation:

$$\hat{U}_{\mathrm{prep}}(\phi_1, \phi_2)|0\rangle = \begin{bmatrix} e^{i\phi_2/2} & 0 \\ 0 & e^{-i\phi_2/2} \end{bmatrix} i \begin{bmatrix} \sin(\phi_1/2) & \cos(\phi_1/2) \\ \cos(\phi_1/2) & -\sin(\phi_1/2) \end{bmatrix} \begin{bmatrix} 1 \\ 0 \end{bmatrix}$$

$$= i \left(e^{i\phi_2/2} \sin(\phi_1/2)|0\rangle + e^{-i\phi_2/2} \cos(\phi_1/2)|1\rangle \right) \tag{2.17}$$

$$\rightarrow |\psi(\phi_1, \phi_2)\rangle_{\mathrm{out}} = \sin(\phi_1/2)|0\rangle + e^{-i\phi_2} \cos(\phi_1/2)|1\rangle, \tag{2.18}$$

where we have neglected the phase $ie^{-i\phi_2/2}$. Equation (2.18) thus parametrizes an arbitrary single-qubit state, up to a global phase. Phase settings to prepare commonly-used single-qubit states are given in the table below.

	$\|0\rangle$	$\|1\rangle$	$\|+\rangle$	$\|-\rangle$	$\|+i\rangle$	$\|-i\rangle$
ϕ_1	π	0	$\pi/2$	$3\pi/2$	$\pi/2$	$\pi/2$
$s\phi_2$	0	0	0	0	$3\pi/2$	$\pi/2$

2.2.6 *Measurement*

By a similar argument, arbitrary single-qubit projective measurements can be performed using an MZI with two phaseshifters ϕ_1, ϕ_2, together with two singl-photon detectors D_0, D_1 (Fig. 2.5b). Each detector projects onto a logical basis state

$$\hat{\Pi}_{D_0} = |0\rangle\langle 0|; \quad \hat{\Pi}_{D_1} = |1\rangle\langle 1|; \quad P(0|\psi) = |\langle 0|\psi\rangle|^2 = \mathrm{Tr}\left[\hat{\rho}\,\hat{\Pi}_{D_0}\right] = 1 - P(1|\psi). \tag{2.19}$$

Assigning eigenvalues of ± 1, the effect of the two detectors together can be written as a projective measurement $\hat{M}$ with spectral decomposition

$$\hat{M} = \sum_i \lambda_i |\lambda_i\rangle\langle\lambda_i| = |0\rangle\langle 0| - |1\rangle\langle 1|, \tag{2.20}$$

which is equivalent to measurement in the z-basis ($\hat{M} = \hat{\sigma}_z$). To measure in a different basis, we apply a unitary rotation $\hat{U}_{\text{meas}}$ to each qubit prior to detection using the MZI shown in Fig. 2.5b. This evolves an input state $|\psi\rangle_{\text{in}}$ as

$$|\psi\rangle_{\text{out}} = \hat{U}_{\text{meas}}(\phi_1, \phi_2)|\psi\rangle_{\text{in}} = -i \begin{bmatrix} \sin(\phi_2/2) & \cos(\phi_2/2) \\ \cos(\phi_2/2) & -\sin(\phi_2/2) \end{bmatrix} \begin{bmatrix} e^{i\phi_1/2} & 0 \\ 0 & e^{-i\phi_1/2} \end{bmatrix} |\psi\rangle_{\text{in}} \tag{2.21}$$

and the overlap between the $|\psi\rangle_{\text{in}}$ and each eigenstate $|\lambda_i\rangle$ of $\hat{\sigma}_z$ becomes $\langle\lambda_i|\psi_{\text{in}}\rangle = \langle\lambda_i|\hat{U}_{\text{meas}}|\psi_{\text{out}}\rangle$. To find the *effective* measurement operator $\hat{M}'(\phi_1, \phi_2)$, we therefore propagate the projectors (2.19) *backwards* through the unitary

$$|\lambda_i'\rangle = \hat{U}^\dagger_{\text{meas}}(\phi_1, \phi_2)|\lambda_i\rangle; \tag{2.22}$$

$$\hat{M}'(\phi_1, \phi_2) = \sum_i \lambda_i |\lambda_i'\rangle\langle\lambda_i'| = \hat{U}^\dagger_{\text{meas}}(\phi_1, \phi_2)\, \hat{\sigma}_z\, \hat{U}_{\text{meas}}(\phi_1, \phi_2). \tag{2.23}$$

By a similar argument to that used in Sect. 2.2.5, $\hat{U}^\dagger_{\text{meas}}$ can map $|0\rangle$ and $|1\rangle$ to any desired eigenstate $|\lambda\rangle$, and $\hat{M}'$ can therefore be made to implement any desired single-qubit projective measurement. Phase settings to measure in the Pauli basis are given in the table below.

	$\hat{\sigma}_x$	$\hat{\sigma}_y$	$\hat{\sigma}_z$
ϕ_1	0	$\pi/2$	0
ϕ_2	$\pi/2$	$3\pi/2$	π

2.2.7 CNOT-MZ Is Universal

The CNOT-MZ can prepare any entangled or separable pure two-qubit state, up to a global phase. To see this, first note that by the Schmidt decomposition [29], any pure two-qubit state can be expressed as an arbitrary superposition of two orthogonal separable states

$$|\Psi_{CT}\rangle = \alpha|0_C 0_T\rangle + \beta|0_C 1_T\rangle + \gamma|1_C 0_T\rangle + \delta|1_C 1_T\rangle \tag{2.24}$$

$$= \sqrt{\lambda}\,|\lambda_C\rangle \otimes |\lambda_T\rangle + \sqrt{1-\lambda}\,|\lambda_C^\perp\rangle \otimes |\lambda_T^\perp\rangle. \tag{2.25}$$

where λ is a real nonnegative number. This immediately implies that the state has six independent real parameters

$$|\Psi_{CT}\rangle = \sqrt{\lambda}\left(\cos\theta_C|0\rangle + e^{i\phi_C}\sin\theta_C|1\rangle\right)\left(\cos\theta_T|0\rangle + e^{i\phi_T}\sin\theta_T|1\rangle\right) \tag{2.26}$$

$$+ e^{i\phi_r}\sqrt{1-\lambda}\left(e^{-i\phi_C}\sin\theta_C|0\rangle - \cos\theta_C|1\rangle\right)\left(e^{-i\phi_T}\sin\theta_T|0\rangle - \cos\theta_T|1\rangle\right), \tag{2.27}$$

up to a global phase. To show that this arbitrary state can be prepared by the CNOT-MZ with $|00\rangle$ as input, we will propagate (2.25) backwards through the circuit. By the same argument given in Sect. 2.2.5, the MZI comprising DCs c_{10} and c_{12}, together with phaseshifters ϕ_5 and ϕ_7, can be configured to map the control qubit into the $|0\rangle$, $|1\rangle$ basis

$$|\Psi'\rangle = \left(\hat{U}^{\dagger}_{\text{meas}}(\phi_5, \phi_7) \otimes \mathbf{1}\right) |\Psi_{CT}\rangle = \sqrt{\lambda}\, |0\rangle \otimes |\lambda_T\rangle + e^{i\phi_r} \sqrt{1-\lambda}\, |1\rangle \otimes |\lambda_T^{\perp}\rangle. \tag{2.28}$$

Propagating backwards through the CNOT-P gate, the target qubit is flipped conditional on the control:

$$|\Psi''\rangle = \hat{U}^{\dagger}_{\text{CNOT}-\text{P}} |\Psi'\rangle = \sqrt{\lambda}\, |0\rangle \otimes |\lambda_T\rangle + e^{i\phi_r} \sqrt{1-\lambda}\, |1\rangle \otimes |\lambda_T\rangle. \tag{2.29}$$

We then use the MZI formed by DCs c_2 and c_4, together with ϕ_2 and ϕ_4, to rotate the target qubit:

$$|\Psi'''\rangle = \left(\mathbf{1} \otimes \hat{U}^{\dagger}_{\text{prep}}(\phi_2, \phi_4)\right) |\Psi''\rangle = \left(\sqrt{\lambda}\, |0\rangle + e^{i\phi_r} \sqrt{1-\lambda}\, |1\rangle\right) \otimes |0\rangle, \tag{2.30}$$

and finally rotate the control, using c_1 and c_3 together with ϕ_1 and ϕ_3

$$|\Psi\rangle_{\text{in}} = \left(\hat{U}^{\dagger}_{\text{prep}}(\phi_1, \phi_3) \otimes \mathbf{1}\right) |\Psi'''\rangle = |00\rangle. \tag{2.31}$$

This capability is used to the fullest extent in Chap. 5 of this thesis.

2.3 Experimental Setup

The full experimental setup is shown schematically in Fig. 2.6. The input and output ports of the CNOT-MZ were butt-coupled to two V-groove fiber arrays, each holding six single-mode optical fibres with 250 μm pitch, to match that of the waveguides. Using an oil-based index-matching fluid at the chip-fibre interface, a fibre-to-fibre coupling efficiency of ~60 % was typically achieved. PMF fibre was used at the input of the chip, so as to preserve indistinguishability of the incoming photon pair, while SMF was employed at the output. The chip die was mounted on a standard PCB, to which the electrodes of each resistive heater were gold-wire bonded. This PCB provides a pinout via two standard 8-pin headers to an 8-channel DC current source.

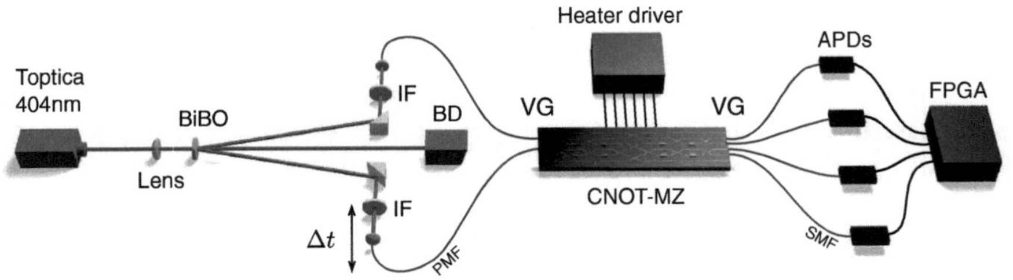

Fig. 2.6 CNOT-MZ experimental setup. A Toptica *iBeam* 404 nm CW laser pumps a BIBO non-linear crystal, cut and phase-matched to generate degenerate 808 nm photon pairs by type-I SPDC. Spectral indistinguishability is optimized using tilted Semrock *Maxline* 3 nm notch interference filters (IF). The pump is absorbed by a beam dump (BD). Photon pairs are coupled in and out of the CNOT-MZ through optical fibre and V-groove fiber arrays (VG). PMF is used at the input, as HOM interference is sensitive to the polarization of incoming photons, while SMF can be used at the output, as the detectors are not strongly polarization-sensitive. A current source connects to resistive heaters onboard the chip via a custom PCB. Four Si-APD single-photon detectors, together with an FPGA, are used to count coincidences at the output of the chip

2.3.1 Photon Pair Source

The CNOT-MZ requires two indistinguishable photons as input. Arguably (see Ref. [30]), the CNOT-P gate does not depend on entanglement from the source—certainly, the Fock state needed to run the gate and encode the control and target qubits, $|1_{V1}1_{V2}\rangle = |VV\rangle$ will not violate a Bell inequality *as-is*, and is not entangled in polarization. This state is naturally generated by postselection on coincidental detection of two photons from the type-I SPDC state (1.168).

The two-photon source used throughout this thesis is shown in Fig. 2.6. A I404 nm CW laser (Toptica *iBeam*) pumps a 2 mm-thick BiBO crystal, cut and phase-matched for type-I SPDC, with a 3° opening angle. Downconverted photon pairs, both of which are vertically polarized, were filtered using 3 nm full-width half-maximum (FWHM) notch interference filters (IFs), and then coupled into PMF using an arrangement of prisms together with 11 mm aspheric lenses. One collection stage was mounted on a motorized linear actuator with micron resolution, allowing the relative arrival time—and thus the temporal distinguishability—of the photon pair to be precisely controlled. Using Perkin-Elmer silicon APD single-photon detectors with a quantum efficiency of ~60 %, we measured a typical single-photon count-rate S of $\sim 1 \times 10^6$ Hz, and a coincidence count-rate C of $\sim 1 \times 10^5$ Hz, implying a collection efficiency of $C/S \approx 10\,\%$.

Photon indistinguishability is a crucial factor for high-fidelity operation of the CNOT-P gate. We first ensured temporal overlap of the downconverted photon pair by matching optical path lengths of the two arms of the source to within the photon coherence length (~500 μm) using the linear actuator, measuring two-photon HOM interference in a fiber-coupled 50:50 BS. In order to optimize the spectral indistinguishability of the photon pair, we measured spectra of down-converted photons

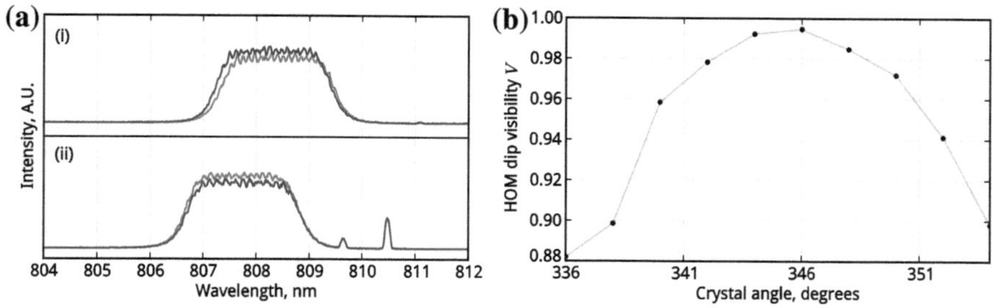

Fig. 2.7 The visibility of the HOM dip is a crucial factor for the performance of the CNOT-P gate. A number of measures were taken to optimize the visibility of quantum interference between photon pairs generated by the type-I source. **a** Experimental data showing the spectra of single photons generated in the two arms of the source (*red*, *blue* respectively). *(i)* Spectra measured prior to optimization of the source. By tilting interference filters placed in each beam, we ensured that photon pairs sent to the CNOT-MZ were maximally spectrally indistinguishable. *(ii)* The small peaks are due to stray light from an LCD computer monitor. **b** HOM visibility measured as a function of BiBO crystal orientation, which affects the polarization and spectral distinguishability of downconverted photon pairs (color online)

in each arm while tilting interference filters, shifting the wavelength of the transmitted band (Fig. 2.7a) and leading to a measurable increase in the visibility of the HOM dip. Finally, we scanned the orientation of the BiBO crystal which affects both pair collection efficiency and polarization distinguishability, further optimizing the visibility of quantum interference (Fig. 2.7b).

2.3.2 *Control, Automation and Readout*

Many of the experiments presented throughout this thesis depend on the ability to perform hundreds or thousands of consecutive measurements, each with different phase settings. As such, it was important that the experimental setup be fully automated. The eight heaters of the CNOT-MZ were driven by a National Instruments digital-to-analog converter (DAC), providing eight computer-controlled voltages in the range [0, 7] V. An eight-channel current amplifier was necessary to satisfy the power draw of the heaters, a total of $\sim$1 W per heater at maximum voltage.

Under typical conditions, when all eight heaters are active, the chip dissipates around $\sim$1 W of heat energy. An experimental difficulty is then presented by the fact that the top surface of the chip, where the heaters and waveguides are located, is raised to a higher temperature than the substrate, leading to thermal expansion and distortion of the chip itself. This leads to movement of the chip facets and decoupling of the waveguides from the V-groove arrays (VGs), as shown in Fig. 2.8c. To solve this issue, we found that the best compromise between coupling efficiency, stability and repeatability was achieved by *pulsing* current to the heaters, with a duty cycle $t_{\text{measure}}/t_{\text{cool}} \sim 5\,\%$. Current was first supplied to the chip for 1 s, allowing the

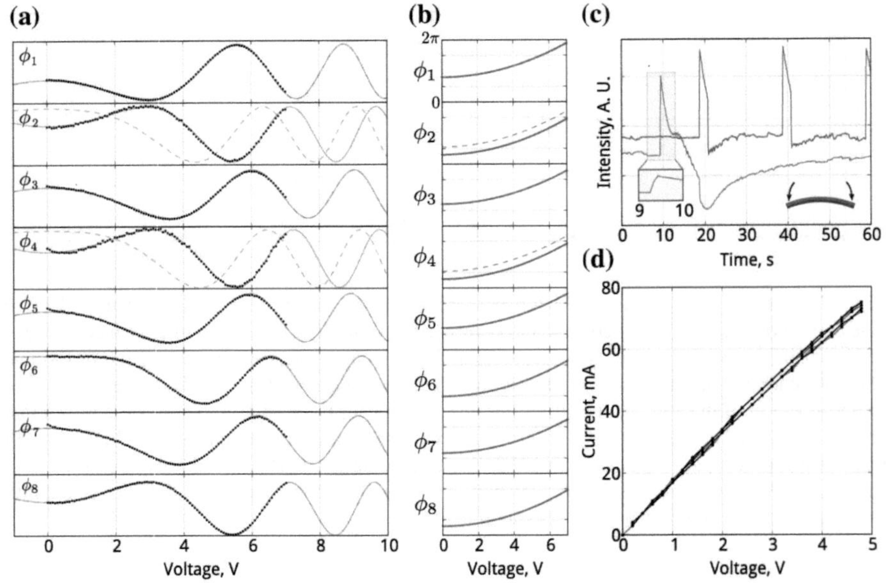

Fig. 2.8 Calibrating the CNOT-MZ. **a** Single-photon interference fringes, measured using heralded single photons from the SPDC source, as a function of resistive heater control voltage. *Black dots* show the experimental data, up to a maximum rated voltage of 7 V. *Blue lines* show a fit to the data, whose parameters completely characterise the phase-voltage relation of each heater. **b** Phase-voltage relations for each thermal phase shifter, based on fit parameters from (**a**). The dominant component is quadratic, $\phi \propto \Delta T \propto P = IV \propto V^2$. **c** Optical intensity measured at the output of the CNOt-MZ, as heaters are switched on and off. The chip deforms under load, resulting in optical decoupling of the V-groove arrays, seen as an immediate dip in intensity as the heater is switched and held on (*red line*). In order to minimize the extent of decoupling we pulse current to each heater, only measuring coincidence events while the heater is switched on (*blue line*). *Inset* zoom showing the response time of the phaseshifter, ~100 ms. **d** Superimposed I-V curves of all eight heaters. The characteristic nonlinearity at high voltage is due to increased resistance of the heating element at high temperatures (color online)

phaseshifter to warm up and stabilize, and was then held on for a further 1 s, while single-photon detection events were measured. The current source was then switched off, allowing the chip to cool for 15 s, after which the cycle was repeated for the next measurement setting.

This decoupling effect was exacerbated by the fact that the fiberglass PCB material, upon which the chip was directly mounted, is a thermal insulator. Ideally, the chip would instead be mounted on a conducting heat sink, or a Peltier-effect thermoelectric cooling system. We expect that this difficulty could be further mitigated using standard chip packaging techniques, in which the VGs are glued directly to the chip facets. Dispensing with the need to periodically cool the chip would lead to an overall improvement in efficiency by a factor of ~20. This would facilitate experiments demanding large numbers of measurements, such as those described in Chap. 5. As with classical classical central processing units (CPUs), heat dissipation will likely remain a significant experimental consideration as the scale and complexity of reconfigurable integrated quantum photonic chips is increased.

The coincidence-counting system was based around a Xilinx *Virtex-5* FPGA. This system was configured to count a specified subset of single detection events and coincidences, with a fixed coincidence window of 5 ns. In all coincidence-counting experiments there is a nonzero probability of detection of temporally distinguishable photons generated in separate downconversion events. These *accidental* coincidences lead to a constant background coincidence rate (5 % of the true count-rate), reducing the apparent visibility (1.134) of quantum interference. In order to correct for this background and obtain a more accurate measure of the performance of the device, during all single-photon measurements presented in this thesis (except those described in Chap. 6 and Sect. 4.5), the background rate of accidental coincidences for each detection pattern was constantly measured and subtracted from the experimental data. This measurement was performed by inserting an electronic delay $\gg 5$ ns between pairs of detectors, and measuring the resulting coincidence count-rate. See Sect. 6.2 for further discussion of correlated single-photon counting systems.

Scripting and control of the experimental setup was performed using the *Python* programming language together with a custom library, qy. More recently, access to the CNOT-MZ has been made available to other researchers and the general public via an open web interface. Further detail regarding scripting and remote automation of theCNOT-MZ is given in Appendix A.

2.3.3 Calibration

Applying a voltage V to the resistive heater of a particular MZI, we obtain a phase shift ϕ. In order to choose the voltage required to apply a desired phase shift at a particular MZI, we must find and invert the phase-voltage relation $\phi(V)$. Since the phaseshift is proportional to the change in temperature of the waveguide material, the phase-voltage relation is approximately quadratic

$$\phi(V, \vec{a}) = a_0 + a_2 V^2 + a_3 V^3 + \text{h.c}; \quad a_3 \ll a_2, \tag{2.32}$$

where $\vec{a}$ are calibration parameters depending on the geometry and fabrication of the heater and surrounding waveguides. Here, a_3 accounts for higher-order effects such as those shown in Fig. 2.8d, and a_0 is the phase in the interferometer at $V = 0$, i.e. when the resistive heater is switched off. Imperfect waveguide geometry, together with imperfections introduced during lithographic fabrication of the heaters themselves, lead to each MZI having a small nonzero value of a_0, which must be individually calibrated. Moreover, small inconsistencies in heater fabrication lead to variance in the values of a_2 and a_3, which also must be individually characterised.

This calibration procedure was accomplished using simple single photon measurements. If bright light or single photons are injected into one port of an MZI, the measured intensity at a given output port is a sinusoidal function of $\phi(V, \vec{a})$,

$$I_{D_0} = I_0 \sin^2(\phi(V, \vec{a})/2)\,; \quad I_{D_1} = I_0 \cos^2(\phi(V, \vec{a})/2)\,. \tag{2.33}$$

Using single photon detectors, we measured fringes of this type for each phaseshifter of the CNOT-MZ, as shown in Fig. 2.8. We fit curves of the form (2.33) to this data with $\vec{a}$ and I_0 as free parameters, thus recovering the unique phase-voltage relation of each heater (Fig. 2.8). By numerically inverting this function, we can find the voltage required to set any desired phase in the interval $[0, 2\pi]$ to any heater on the CNOT-MZ.

Owing to the geometry of the device, it is not always possible to directly inject light into a single input port of a particular MZI under test. Moreover, the contrast of the measured fringe is sometimes dependent on the (initially unknown) phase inside another interferometer: an example of such an interdependence is seen between phaseshifters ϕ_2 and ϕ_4. As a result, the full calibration procedure had to be completed in two stages. We first measured "rough" fringes with only a single resistive heater active at any given time. Approximate information obtained from these measurements was then used to take full-contrast fringes (Fig. 2.9) in a second pass, activating multiple phaseshifters at once to optimize contrast and signal-to-noise ratio. We expect that such techniques will need to be considerably refined as the scale and complexity of reconfigurable quantum photonic chips is increased. Progress on automatic calibration and characterization of such devices was recently described by Li et al. [31].

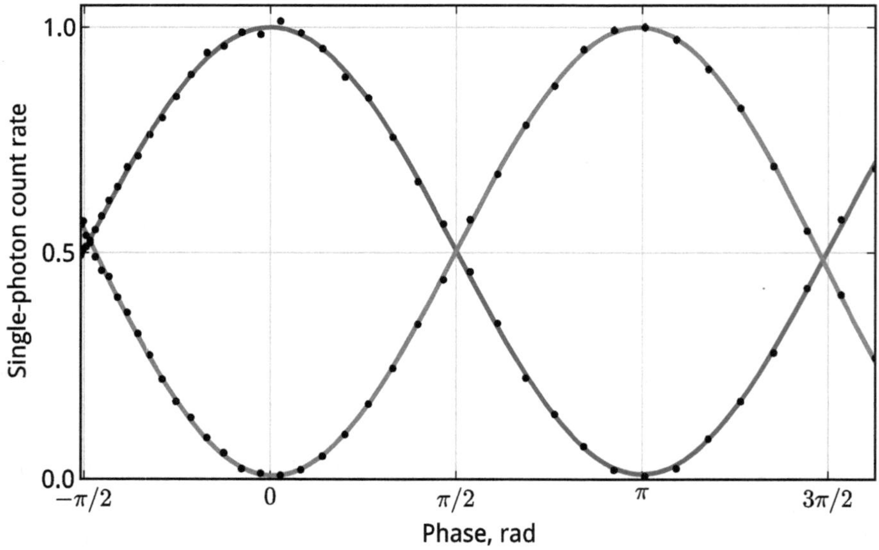

Fig. 2.9 Single-photon interference fringe, measured at the two outputs of a single MZI on the CNOT-MZ. Experimental data are presented as *black circles*, *solid lines* show fits to the theory. *Error bars*, which assume Poissonian statistics, are too small to draw (color online)

As shown in Sect. 1.5.1, uncontrolled polarization rotations in the waveguide, or coupling to higher-order spatial guided modes, would give rise to reduced contrast in these single-photon fringes, as would thermal or electric fluctuations (e.g. DAC noise) in the phase shifter. These effects would reduce the fidelity with which single-qubit states and measurements can be implemented, and would to all intents and purposes resemble decoherence of the photonic qubit,[2] adding unwanted mixture to the state. High-contrast single-photon fringes are therefore a good indicator of the quality and single-mode operation of the waveguides, and are a prerequisite for high-fidelity quantum operations. We measured an average contrast over all eight fringes of $\bar{C} = 0.988 \pm 0.008$. From these fringes, we estimated the average experimental accuracy in phase to be $\delta_\phi \sim 0.05$ rad. We did not find any significant evidence of thermal cross-talk between phaseshifters.

2.4 On-Chip Quantum Interference

In addition to high-fidelity classical interference, as demonstrated in Fig. 2.9, the basic mechanism of the CNOT-P gate relies on high-fidelity quantum interference. The same effects that would give rise to reduced contrast of single-photon interference would also render photon pairs distinguishable, reducing the visibility of the HOM dip and thus having a detrimental effect on the performance of the entangling gate.

In order to accurately assess the visibility of HOM interference supported by the CNOT-MZ, we first set $\phi_1 = \pi/2$, rendering the interferometer formed by DCs c_1 and c_3 (Fig. 2.1) equivalent to a 50:50 BS. Injecting single photon pairs from the source into waveguides w_2 and w_3, we measured the coincidence count-rate $C(\Delta t)$ at output ports w_1 and w_4, as a function of the linear actuator position—corresponding to a difference Δt in the relative arrival time of the photon pair. The resulting HOM dip is shown in Fig. 2.10.

The shape of the HOM dip is given by a convolution of the wavepacket of down-converted photons and the top-hat profile of the interference filters. It it therefore well-approximated by a function consisting of Gaussian and sinc terms, together with a linear term to account for decoupling of the source as the actuator is moved:

$$C(\Delta t) \approx (a_1 \Delta t + a_2)\left[1 - V \exp\left(-\frac{(\Delta t - a_3)^2}{2a_4^2}\right)\text{sinc}\,(a_5 \Delta t + a_6)\right] \tag{2.34}$$

where $\vec{a}$ are free parameters, and V is the visibility of quantum interference (1.134). Fitting this curve to the data shown in Fig. 2.10, we found $V = 0.978 \pm 0.007$, taking into account the measured rate of accidental coincidences. Here uncertainty was estimated using a Monte-Carlo technique, assuming Poissonian statistics.

[2]See Sect. 1.6.1.

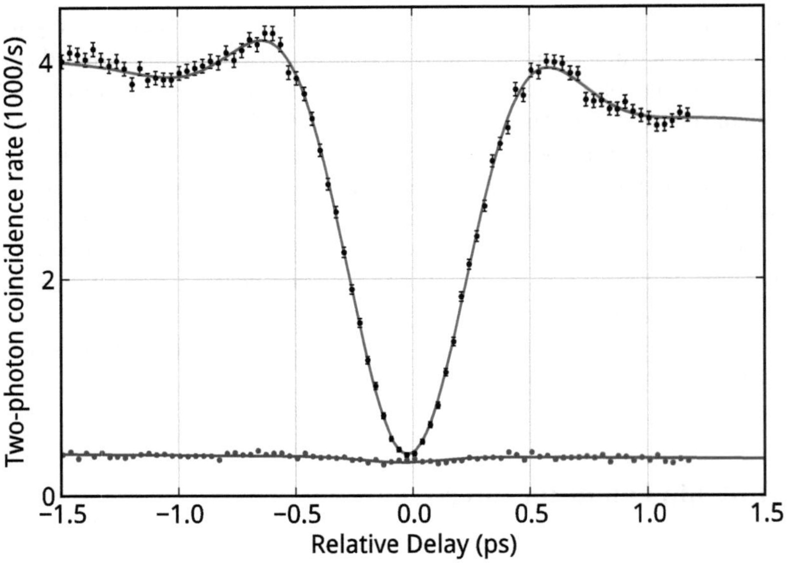

Fig. 2.10 A HOM dip, measured using a single MZI on the CNOT-MZ as a 50:50 BS, as a function of a relative delay between photon pair arrival times, controlled using the linear actuator shown in Fig. 2.6. Measured two-photon coincidence count-rates are shown as *black dots*. The *red line* shows a fit to this data comprising Gaussian, sinc, and linear terms (2.34). The *blue line* shows a fit to the measured rate of accidental coincidences, with Gaussian and linear components. *Error bars* assume Poissonian statistics (color online)

2.5 Randomized Benchmarking

Having calibrated each phaseshifter and observed high-visibility quantum interference in the CNOT-MZ, we then used a randomized benchmarking technique to to characterise the operational real-world performance of the device, across the full parameter space. We cannot expect to test every possible configuration of all eight phase shifters. Instead, we checked performance for a large number of configurations sampled uniformly at random from the full 8-dimensional parameter space of the chip. A somewhat similar randomized approach to global characterization of quantum gate operations has been proposed by Knill [32].

We first chose 1000 random vectors $\vec{\phi}_j$ representing possible configurations of the device

$$\vec{\phi}_j = \left[\phi_{1,j}, \phi_{2,j}, ..., \phi_{8,j}\right]; \quad 0 \leq \phi_{ij} \leq 2\pi. \tag{2.35}$$

Injecting indistinguishable photon pairs into waveguides w_2 and w_4, we encoded the logical qubit state $|00\rangle$ at the input of the device. For each configuration $\vec{\phi}_j$, we then measured coincidence count rates at the output, postselecting on the 2-qubit subspace of detection patterns

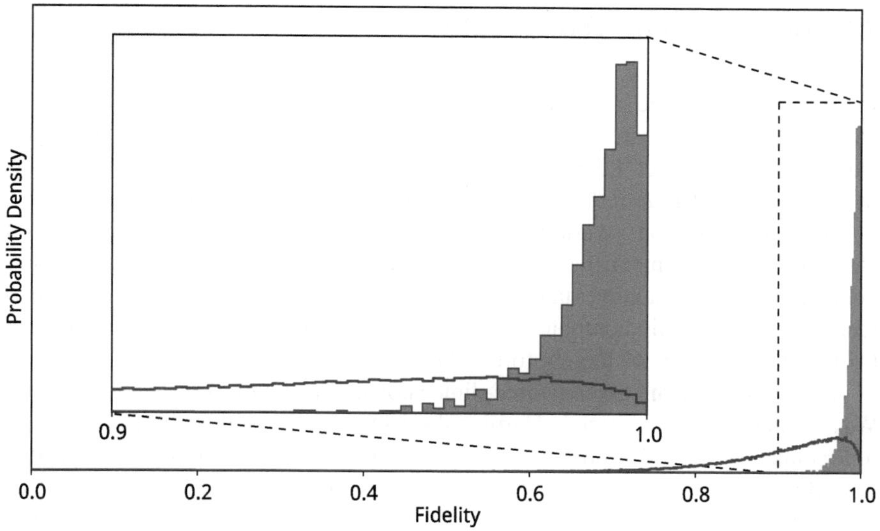

Fig. 2.11 Randomized benchmarking of the CNOT-MZ. The *histogram* shows the distribution of statistical fidelity $F(\vec{P}, \vec{P}')$ between measured coincidence count-rates $\vec{C} \approx C_0 \vec{P}$ and those predicted by an ideal theoretical model $\vec{P}'$, over 995 randomly-chosen phase settings $\vec{\phi}_j$. 96 % of phase settings produced statistics corresponding with theory to $F > 0.97$. The *red line* shows the expected distribution for a device whose output is completely uncorrelated with the desired behaviour, i.e. a white noise source (color online)

$$\vec{C}_j = \left[C_{00,j},\ C_{01,j},\ C_{10,j},\ C_{11,j}\right] \approx \left(\sum_i C_{ij}\right)\vec{P}_j. \tag{2.36}$$

Using an idealized numerical model of the device, assuming unit visibility of quantum interference and perfect fabrication, we then calculated the ideal probability distribution $\vec{P}'_j$ for each configuration of phases. The experimental setup would ideally exactly reproduce the theoretical prediction, $\vec{P}'_j = \vec{P}_j$. We characterised the discrepancy between the performance of the CNOT-MZ and theoretical predictions using the *statistical fidelity* $F(P, P') = \sum_i \sqrt{P_i \cdot P'_i}$. The measured statistical distribution of these fidelities over 995 random configurations[3] is shown in Fig. 2.11. The average fidelity across all configurations was measured to be 0.990 ± 0.009 with 96 % of configurations producing photon statistics with $F > 0.97$.

This result depends on simultaneous high fidelity quantum and classical interference, as well as accurate joint control of all eight phase controllers. Poor performance of any of these component parts would result in lower fidelity output for some subset of configurations. The fact that we see good fidelity over many random trials allows us to progress to more rigorous and sophisticated tests, described in the remainder of this chapter.

[3]Five measurement outcomes were deemed to be spurious due to detectable experimental error.

2.6 Quantum State Tomography

In order to characterize states generated by the experimental apparatus, we will often make use of simple witnesses and metrics such as Bell-CHSH, concurrence, etc. (see, for example, Sect. 1.3.8). However, the most complete information is encoded in the density matrix $\hat{\rho}$ of the experimental state itself. We performed quantum state tomography (QST) on a variety of states generated by the CNOT-MZ, using on-chip MZIs to implement the requisite measurements and reconstruct $\hat{\rho}$. Previous demonstrations of quantum state tomography in integrated photonics have not used reconfigurable on-chip components to implement the different settings required for QST. In this analysis we largely follow James et al. [33].

Imagine that we are given a three-dimensional object with some complex shape. We are interested in completely learning the 3-D geometry of this object. It is natural to first take a fixed viewpoint, projecting the 3-D structure of the object in question onto the 2-D retina of the eye. With this information in mind, we then rotate the object, and make a second projective measurement. Again we rotate, and project, and rotate and so on, until after some sufficient number of measurements we can completely reconstruct the object in the abstract 3-D space of the mind's eye. Medical imaging techniques such as X-ray computed tomography (CT) and magnetic resonance imaging (MRI) make use of this method.

An analogous task exists for quantum states. In experiments, we are often presented with a device or source which generates a quantum state $\hat{\rho}$ which is partially or entirely unknown or untrusted. Using QST [29, 33], the full density matrix can be approximately (and in some cases exactly) reconstructed, by making an appropriate set of projective measurements on a number of copies of the state $\hat{\rho}$. The origins of the technique arguably lie with Stokes [34], who described a method to fully reconstruct the polarization of a beam of light based on simple measurements.

QST, while closely analogous to classical tomography, is distinguished by the fact that, for quantum systems, measurement necessarily changes the state of the object under test. Therefore, we cannot always perform consecutive measurements $\hat{\tau}_i$ on a *single copy* of a quantum state $\hat{\rho}$ and expect to accurately recover the expectation values $\langle\hat{\tau}_i\rangle = Tr[\hat{\tau}_i\hat{\rho}]$. This notion is captured in Heisenberg's uncertainty principle and is a direct result of the No-Cloning theorem (see Sect. 1.3.4). Since the observer cannot clone the system without prior knowledge of $\hat{\rho}$, we usually consider tomographic situations where a "black box" device repeatedly outputs $\hat{\rho}$ on-demand, and consecutive measurements are evaluated on copies of the state generated in this way.

In this discussion we will consider a system of n qubits, however the analysis easily extends to higher-dimensional systems [35]. A general n-qubit mixed state can be written as

$$\hat{\rho} = \frac{1}{2^n} \sum_{i_1,i_2...i_n=0}^{3} S_{i_1,i_2...i_n} \hat{\sigma}_{i_1} \otimes \hat{\sigma}_{i_2} \otimes \cdots \otimes \hat{\sigma}_{i_n} \tag{2.37}$$

where $\hat{\sigma}_i$ are the Pauli matrices and $\{S_{i_1,i_2...i_n}\} \equiv \vec{S}$ are the *Stokes parameters*, 4^n real numbers which together completely and uniquely characterise $\hat{\rho}$. Complete knowledge of $\vec{S}$ amounts to complete knowledge of the physical state of the system. Normalization imposes the condition that $S_{0,0...0} = 1$, leaving $4^n - 1$ real parameters to be estimated.

The set of n-qubit measurement operators $\{\hat{\tau}_i\}$ used for QST is referred to as the *quorum*. A remarkable property of QST is that regardless of the degree of entanglement of $\hat{\rho}$, there is no need to measure in entangled bases. Although entangled measurements have advantages for certain tomographic applications [36], experimentally it is often dramatically more convenient to measure in a separable basis. $4^n - 1$ local measurement operators of the form $\hat{\tau}_i = \hat{\tau}_{i_1} \otimes \hat{\tau}_{i_2} \otimes \cdots \otimes \hat{\tau}_{i_n}$ therefore suffice for the reconstruction of any $\hat{\rho}$, where $\hat{\tau}_{i_j}$ is a 2×2 single-qubit measurement operator on the jth qubit. When examining a classical 3D object, if we always observe the object from one angle, changing only our distance from the sample, we will not obtain full information of its shape. Similarly, for complete reconstruction of an unknown $\hat{\rho}$ each measurement in the quorum must be linearly independent from all others, i.e. a given $\hat{\tau}_i$ cannot be written as a linear sum over the remaining $\{\hat{\tau}_{i' \neq i}\}$.

Experimentally, we measure the expectation values

$$\vec{T} \equiv \{T_i\}; \quad T_i = \sum_j \frac{\lambda_{ij} c_{ij}}{C_i} \mathrm{Tr}\left(\hat{\rho}\, |\lambda_{ij}\rangle\langle\lambda_{ij}|\right) \approx \langle\hat{\tau}_i\rangle = \mathrm{Tr}\left[\hat{\rho}\hat{\tau}_i\right] \tag{2.38}$$

over the quorum $\{\hat{\tau}_i\}$. where $\{|\lambda_{ij}\rangle, \lambda_{ij}\}$ are the eigenstates and eigenvalues of the experimental measurement measurement operator $\tilde{\tau}_{ij} \approx \hat{\tau}_{ij}$, c_{ij} is the count-rate corresponding to detection of $|\lambda_{ij}\rangle$, and $C_i = \sum_j c_{ij}$ is the total number of detection events for a particular measurement setting. Having obtained $\vec{T}$, the experimental density matrix is typically reconstructed using one of two standard approaches: *linear* or *maximum-likelihood* reconstruction.

2.6.1 Linear Reconstruction

Consider the choice of quorum

$$\hat{\tau}_i = \hat{\sigma}_{i_1} \otimes \hat{\sigma}_{i_2} \otimes \cdots \otimes \hat{\sigma}_{i_n}, \tag{2.39}$$

where σ_i are the usual Pauli matrices. It is easy to see from (2.37) that for this quorum, under ideal experimental conditions, $\vec{T} = \vec{S}$—in which case $\hat{\rho}$ can be simply reconstructed by evaluation of the sum in (2.37). The simplicity of this reconstruction motivates (2.39) as the quorum of choice in many experimental implementations of QST. However, it is not necessary to choose (2.39) and there are sometimes experimental reasons[4] to make a different choice. In particular it is not necessary for the eigenstates of $\hat{\tau}_i$ to be orthogonal. In order to accommodate more general quora in this analysis, we can write the system of simultaneous equations relating $\vec{T}$ and $\vec{S}$ as $\vec{T} = Q\vec{S}$ where Q is a change-of-basis matrix with entries

$$Q_{i,j} = \frac{1}{2^n}\mathrm{Tr}\left[\hat{\tau}_i\hat{\sigma}_j\right]. \tag{2.40}$$

This allows $\vec{T}$—the experimental data—to be converted to $\vec{S}$ by linear inversion of Q, which is guaranteed to be possible because $\hat{\tau}_i$ and $\hat{\sigma}_j$ are both linearly independent. Once this is done, reconstruction of $\hat{\rho}$ is a simple matter of evaluating (2.37).

2.6.2 *Maximum Likelihood Quantum State Tomography*

Linear reconstruction as described above is attractive because of its simplicity. However in real experiments, finite statistics, errors in the implementation of $\hat{\tau}$ [37], and detection errors, for example dark counts (Sect. 1.6.4), all give rise to imperfection and noise in $\vec{T}$, resulting in a discrepancy between the true state of the system $\hat{\rho}$ and the reconstructed image $\hat{\rho}_r$. Importantly, linear reconstruction can yield instances of $\hat{\rho}_r$ which are not physical, i.e. where one or more of the conditions that $\hat{\rho}_r$ should be trace-one, positive-semidefinite, and Hermitian (see Sect. 1.3.6) are not met.

When $\hat{\rho}_r$ is not physical, we cannot confidently apply standard measures to estimate its properties—for example by computing the quantum state fidelity with respect to an ideal state. As a result, *maximum-likelihood quantum state tomography* [33] was developed to guarantee physicality in reconstructed density matrices. This is accomplished by use of numerical optimization to maximize, over the space of all physical density matrices, a likelihood function describing the probability that a particular $\hat{\rho}_r$ gave rise to the experimental data. The parametrization of this space can be achieved using the following form, which is positive-semidefinite Hermitian and normalized by construction:

$$\hat{\rho}_p\left(\vec{t}\right) = \frac{\hat{g}(\vec{t})\hat{g}(\vec{t})^\dagger}{\mathrm{Tr}\left[\hat{g}(\vec{t})\hat{g}(\vec{t})^\dagger\right]}. \tag{2.41}$$

where $\vec{t}$ is a vector of 4^n real parameters and $\hat{g}$ is a $2^n \times 2^n$ complex matrix.[5] Rather than maximising the likelihood, we can instead minimize the least-squares cost function

[4] See for example Ref. [37], where this problem is addressed for polarization-encoded qubits.

[5] There are many ways to parametrize $\hat{g}$ in terms of $\vec{t}$. The only condition is that $\hat{\rho}_p(\vec{t})$ spans the entire Hilbert space. See [33] for one example.

$$\Gamma\left(\vec{t}\right)=\sum_{i}\frac{\left(\mathrm{Tr}\left[\hat{\rho}_{p}\left(\vec{t}\right)\hat{\tau}_{i}\right]-T_{i}\right)^{2}}{2\mathrm{Tr}\left[\hat{\rho}_{p}\left(\vec{t}\right)\hat{\tau}_{i}\right]},\tag{2.42}$$

with respect to $\vec{t}$. This minimization thus yields a description of the state, $\hat{\rho}_p(\vec{t}_{\max})$, most likely to have generated the experimental data.

In (2.42) it is sufficient to iterate over a minimal set of $4^n - 1$ projective measurements. Although this is experimentally the least costly option, it can be advantageous to include an *over-complete* quorum. Depending on the particulars of the experiment, we can use an arbitrary number of measurement outcomes, over and above the minimal set, without any modification of (2.42). This has the advantage of improved resilience to measurement error and spurious measurement outcomes, with the result that $\hat{\rho}_r$ gives a better approximation to the true state of the system $\hat{\rho}$ (see [37]).

The numerical optimization task of finding the maximum-likelihood state is unsurprisingly computationally demanding, working as it must over $4^n - 1$ parameters. This is compounded by the problem of estimating error bars on quantities computed from the reconstructed state, which—due to the nonlinear, algorithmic nature of the reconstruction process—is typically achieved through a Monte-Carlo approach, requiring on the order of 100 repeated trials of the optimization process. For single-qubit states this can be achieved in an acceptable time using high-level interfaces to general-purpose Nelder-Mead simplex algorithms such as `fminsearch` in Matlab and `scipy.optimize.fmin` in Python. However, for larger systems these functions become unacceptably slow.

It turns out that maximum-likelihood estimation can instead be written as a *semi-definite programme*, a particular class of optimization problems dealing with linear functions of positive semidefinite Hermitian matrices: i.e. density operators. By exploiting this knowledge along with the fact that the function (2.42) is *convex*—it has at most one minimum point—we can solve the optimization problem in much less time with respect to general-purpose methods.

It should be emphasised that although general-purpose QST can be made tractable for small systems (on the order of tens of qubits) [38], it is intrinsically exponentially hard to learn or even represent an unknown n-qubit state. When we come to build large-scale quantum computers with thousands or millions of physical qubits, it will not be possible to learn the full state of the system at any point. Various methods have been developed in order to mitigate this problem, many of which make use of prior knowledge or reasonable assumptions on the state to make the representation and tomography efficient. Significant examples include QST by *compressed sensing* [39], which provides a logarithmic speedup with respect to full QST by assuming that the state is relatively pure and therefore sparse in some basis, and *matrix product state* methods [40], which also provide a logarithmic speedup by assuming that the state is constructed by means of a particular sequence of entangling operations between small numbers of adjacent qubits. However, it remains to be seen how well these techniques perform when applied to the diagnosis of imperfection in a real quantum computer, and scalable validation and verification of quantum states remains a topic of considerable interest and urgency. See Sect. 6.3.5 for a discussion of these topics outside the qubit encoding.

2.6.3 On-Chip Quantum State Tomography

Throughout this thesis, we make use of QST to characterize the quality of states generated by the CNOT-MZ. As a first demonstration, we prepared and measured each of the four canonical Bell states (1.38). These states, being maximally entangled, provide a particularly rigorous test of the performance of the CNOT-P gate.

Setting appropriate voltages to phaseshifters ϕ_{1-4} as described in Sect. 2.2.5, we prepared the separable superposition states

$$|+\rangle_C \otimes |0\rangle_T, \quad |+\rangle_C \otimes |1\rangle_T, \quad |-\rangle_C \otimes |0\rangle_T, \quad |-\rangle_C \otimes |1\rangle_T \tag{2.43}$$

at the input of the CNOT-P gate. The corresponding Bell states ($|\Phi^{\pm}\rangle$ and $|\Psi^{\pm}\rangle$ respectively) are then ideally produced at the output.

For each input state, phase shifters ϕ_{5-8} were then used to implement the quorum of 16 measurement settings required to reconstruct the density operator of the state. Since we collect statistics for all four logical outputs of the device simultaneously, is straightforward to implement an over-complete quorum

$$\hat{\tau}_i = |C_i\rangle\langle C_i| \otimes |T_i\rangle\langle T_i| \tag{2.44}$$

over all combinations of $|C_i\rangle, |T_i\rangle \in \{|0\rangle, |1\rangle, |+\rangle, |-\rangle, |+i\rangle, |-i\rangle\}$. The measured density matrices of all four Bell states are shown in Fig. 2.12, with *quantum state fidelities* [29]

$$F = \left(Tr\sqrt{\sqrt{\rho_{th}}\rho_{exp}\sqrt{\rho_{th}}} \right)^2 \tag{2.45}$$

of 0.947 ± 0.002, 0.945 ± 0.002, 0.933 ± 0.002, and 0.885 ± 0.002 respectively.

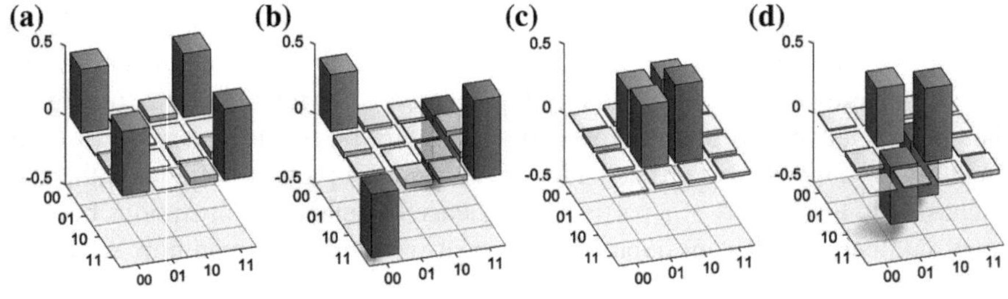

Fig. 2.12 On-chip quantum state tomography. Density matrices of the Bell states **a** $|\Phi^+\rangle$, **b** $|\Phi^-\rangle$, **c** $|\Psi^+\rangle$ and **d** $|\Psi^-\rangle$, generated and characterized on-chip. Imaginary parts are not shown

A discussion of sources of error in the CNOT-MZ, to which we attribute the infidelity seen here, is given in Sect. 2.10.

2.7 Quantum Process Tomography

QST allows us to obtain complete information about the *output* of a black-box source of quantum states. Often, we are also interested in devices which transform an arbitrary input state, where we would like to learn the a priori unknown relationship between input and output states of the device [41].

While many of the errors which arise in LOQC are described by unitary operators,[6] in order to completely describe an arbitrary black-box device it is necessary to account for processes which do not preserve the purity or orthogonality of their input states. This can occur if the system couples to unknown environmental degrees of freedom, which are traced over in the final measurement. Any black box of this type can be completely and uniquely characterised by a completely positive map $\mathcal{E}$. This operator describes the effect of the device on an input state $\hat{\rho}_{in}$,

$$\hat{\rho}_{out} = \mathcal{E}\left(\hat{\rho}_{in}\right) = \sum_i \hat{A}_i \hat{\rho}_{in} \hat{A}_i^\dagger \tag{2.46}$$

where $\hat{A}_i$ are a set of operators acting on the Hilbert space of $\hat{\rho}$. In order to connect this theoretical description with experiment it is helpful to re-write

$$\hat{A}_i = \sum_j a_{i,j} \bar{A}_j \tag{2.47}$$

where $\bar{A}_j$ are the Kraus operators, which are fixed and independent of $\mathcal{E}$. $\bar{A}_j$ satisfy $Tr(\bar{A}_j^\dagger \bar{A}_k) \sim \delta_{j,k}$ and $\sum_j \bar{A}_j^\dagger \bar{A}_j = I$. For qubit systems the Kraus operators are typically chosen as tensor products of Pauli matrices, $\bar{A}_j = \hat{\sigma}_{j_0} \otimes \hat{\sigma}_{j_1} \otimes \cdots \otimes \hat{\sigma}_{j_n}$. The quantum operation can then be completely and uniquely characterised by the *process matrix* $\chi_{m,n} \equiv \sum_i a_{i,m} a_{i,n}^*$, a matrix of 2^{2n} complex numbers with $2^{4n} - 2^{2n}$ free parameters, which relates $\hat{\rho}_{out}$ to $\hat{\rho}_{in}$ as

$$\hat{\rho}_{out} = \mathcal{E}\left(\hat{\rho}_{\text{in}}\right) = \sum_{m,n} \chi_{m,n} \bar{A}_m \hat{\rho}_{in} \bar{A}_n^\dagger. \tag{2.48}$$

The task of quantum process tomography (QPT) is then to estimate χ. For an input state $\hat{\rho}_{\text{in}}$, the probability that the output state of the device is detected in a state $\hat{\tau}_k$ is given by

$$P_{jk} = \text{Tr}\left[\hat{\tau}_k \, \hat{\rho}_{\text{out}}^j\right] = \text{Tr}\left[\hat{\tau}_k \, \mathcal{E}\left(\hat{\rho}_{\text{in}}^j\right)\right]. \tag{2.49}$$

[6] For example, errors in BS reflectivity.

In order to obtain sufficient information to fully reconstruct $\mathcal{E}$ for an arbitrary device, we follow a procedure which is equivalent to full QST of $\hat{\rho}^j_{\text{out}}$, for a complete or over-complete set of linearly independent $\hat{\rho}^j_{\text{in}}$—that is, $\hat{\rho}^j_{\text{in}}$ should at least form a basis for the Hilbert space upon which $\mathcal{E}$ acts. Experimentally, we measure count rates

$$n_{jk} \approx P_{jk} \sum_j n_j = P_{jk}\mathcal{N}; \quad \tilde{P}_{jk} \equiv \frac{n_{jk}}{\mathcal{N}} \tag{2.50}$$

for every possible combination over a quorum of at least $4^n - 1$ input states $\hat{\rho}^j_{\text{in}}$ and $4^n - 1$ measurement settings $\hat{\tau}_k$.

Having acquired this data, we must then reconstruct χ. Although linear reconstruction techniques exist, they suffer the same issues as linear QST: namely, experimental imperfection and finite statistics can lead to a reconstructed process matrix which is unphysical, precluding comparison with standard metrics. As a result, experimental QST is usually performed using a maximum-likelihood reconstruction technique. As with maximum-likelihood QST, we first choose a parametrization of χ which enforces physicality. Since the process matrix is subject to the same physical constraints as a density matrix (both are normalized, Hermitian, positive-semidefinite square matrices), we use a similar parametrization:

$$\mathcal{E}(\vec{t}) \leftrightarrow \tilde{\chi}\left(\vec{t}\right) = \frac{\hat{g}\left(\vec{t}\right)\hat{g}\left(\vec{t}\right)^\dagger}{\text{Tr}\left[\hat{g}\left(\vec{t}\right)\hat{g}\left(\vec{t}\right)^\dagger\right]}. \tag{2.51}$$

We then minimize the cost function [41], constituting a least-squares difference between the observed data and that predicted by theory, with respect to $\vec{t}$:

$$\Gamma(\vec{t}) = \sum_{jk} \frac{\left(\tilde{P}_{jk} - \text{Tr}\left[\hat{\tau}_k\,\mathcal{E}(\vec{t}\hat{\rho}^j_{\text{in}})\right]\right)^2}{2\text{Tr}\left[\hat{\tau}_k\,\mathcal{E}(\vec{t}\hat{\rho}^j_{\text{in}})\right]}. \tag{2.52}$$

Fortunately, this problem can be converted into a semidefinite program [37, 42], allowing the used of convex optimization algorithms which can greatly accelerate the numerical optimization procedure.

2.7.1 On-Chip Quantum Process Tomography

We used the state preparation and measurement stages of the CNOT-MZ to perform full QPT of the CNOT-P gate. This test completely and uniquely characterizes the CNOT-P gate itself, providing full information on the quality of our implementation. In addition, the QPT protocol places stringent demands on the performance of the reconfigurable components of the chip: even if the CNOT-P were perfect, errors in state preparation and measurement would lead to recovery of a flawed process matrix. Moreover, QPT of a 2-qubit gate requires 256 measurements, and is particularly demanding in terms of repeatability and stability of the experimental setup.

Setting appropriate voltages to phase shifters ϕ_{1-4} as described in Sect. 2.2.5, we prepared 16 separable, linearly independent input states

$$\rho_{\text{in}}^{j} = |\Psi_j\rangle\langle\Psi_j| \,; \quad |\Psi_j\rangle = |C_j\rangle \otimes |T_j\rangle \,; \quad |\psi\rangle \in \{|0\rangle, |1\rangle, |+\rangle, |+i\rangle| - i\rangle\}. \tag{2.53}$$

For each ρ_{in}^{j}, the output state of the CNOT-P gate was measured and reconstructed by QST as before, using phase shifters ϕ_{5-8} to perform each of the 16 measurements. These density matrices are shown together with ideal states in Fig. 2.13a.

The process matrix χ was then reconstructed according to the maximum likelihood technique previously described. The experimentally measured process matrix is shown together with the theoretical ideal matrix χ_{ideal} in Fig. 2.13b–d. For clarity, the experimental matrix has been rotated through a local two-qubit unitary which maps CNOT-P to CNOT (see Sect. 2.2.4). The *process fidelity* [29]

$$F_P = \text{Tr}(\chi_{\text{ideal}}\chi_{\text{exp}}) \tag{2.54}$$

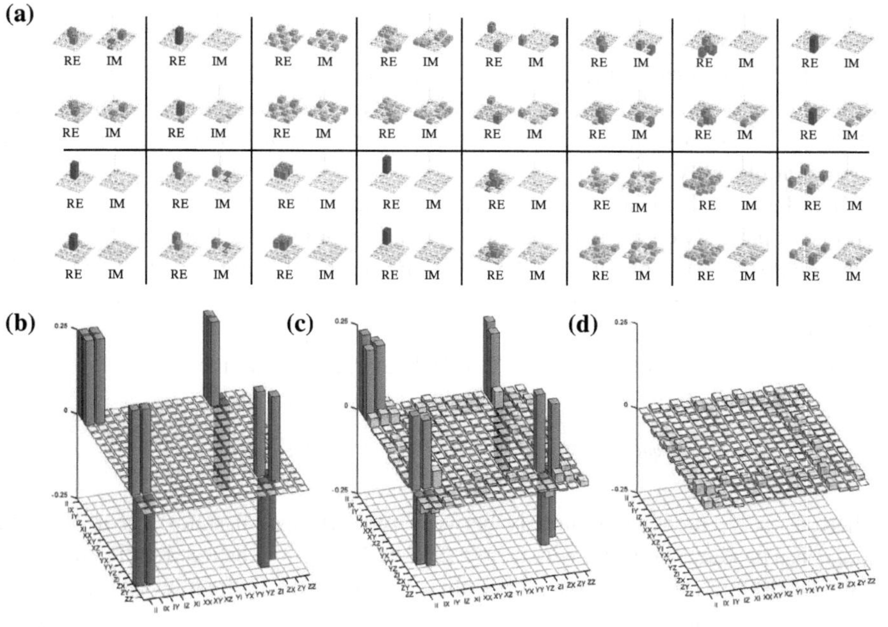

Fig. 2.13 Quantum process tomography of a maximally entangling gate. **a** Ideal and experimental output states of the CNOT-P gate, for a complete set of linearly independent input states. **b** Ideal process matrix of the CNOT gate. The imaginary part is zero everywhere. **c** Real and **d** imaginary parts of the measured process matrix of the CNOT-P device, after a local rotation to permit comparison with the canonical CNOT gate

between the reconstructed process and the ideal CNOT operation was found to be 0.841 ± 0.002. This is comparable with the process fidelity of 0.87 previously measured using an equivalent bulk-optical circuit [27]. The average fidelity [43], defined as the state fidelity between actual and ideal output states averaged over all possible input states, is 0.873 ± 0.001. Here error was determined by a Monte-Carlo approach, assuming Poissonian photon statistics. Sources of error contributing to this sub-unit process fidelity are discussed in Sect. 2.10.

2.8 Bell Inequality Manifold

Having shown that the CNOT-MZ can prepare maximally entangled states, we now demonstrate that these states are nonlocal. As previously discussed, the Bell-CHSH test (Sect. 1.3.8) provides a particularly rigorous criterion for a source of entanglement. In particular, only a subset of the most strongly entangled states can generate nonlocal statistics in a CHSH test. As such, CHSH is important not only as a fundamental test of foundational quantum theory, but also as a measure of the operational performance of quantum technologies and devices.

In the context of the CNOT-MZ, all local realistic models demand that

$$|S| = |\langle \hat{C}_1 \hat{T}_1 \rangle + \langle \hat{C}_1 \hat{T}_2 \rangle + \langle \hat{C}_2 \hat{T}_1 \rangle - \langle \hat{C}_2 \hat{T}_2 \rangle| \leq 2 \tag{2.55}$$

where $\hat{C}_i$, $\hat{T}_j$ are measurement operators on the control and target qubits respectively. If these qubits are entangled, this inequality can be violated up to a maximum value of $|S| = 2\sqrt{2}$—in which case we say that we detect nonlocal statistics, or that we "obtain nonlocality".

In order to further test the reconfigurability of the CNOT-MZ, we measured S over a range of partially entangled states, using a variety of measurement settings. Even if the state $|\Psi(\phi_{1-4})\rangle$ generated by the CNOT-P is maximally entangled, (2.55) is only violated for a subset of measurement settings. See Sect. 4.2 for further discussion of this point.

We used ϕ_{1-4}, together with the CNOT-P gate, to prepare the state

$$|\psi_{out}\rangle = \frac{1}{2\sqrt{2}}[(1 - e^{i\alpha})|00\rangle + (1 + e^{i\alpha})|11\rangle], \tag{2.56}$$

where $\alpha = \phi_1$ tunes continuously between two orthogonal states: for $\alpha = 0, \pi$, $|\psi_{out}\rangle$ is a product state, and with $\alpha = \pi/2, 3\pi/2$, $|\psi_{out}\rangle$ is the maximally entangled state $\frac{1}{\sqrt{2}}(|00\rangle \pm i|11\rangle)$ (up to a global phase). Scanning α in the interval $[0, 2\pi]$, we pass through a continuum of partially entangled states. In order to evaluate S, we used phaseshifters ϕ_{5-8} to implement four two-qubit measurements on the state emerging from the CNOT-P gate. While Alice's measurement settings ($\phi_6 \in \{\pi/4, -\pi/4\}$) were fixed, Bob's measurement operators were continuously rotated in the real plane of the Bloch sphere, with $\phi_8 \in \{\beta, \beta + \pi/2\}$. We measured $S(\alpha, \beta)$ for $\alpha \in [0, 2\pi]$

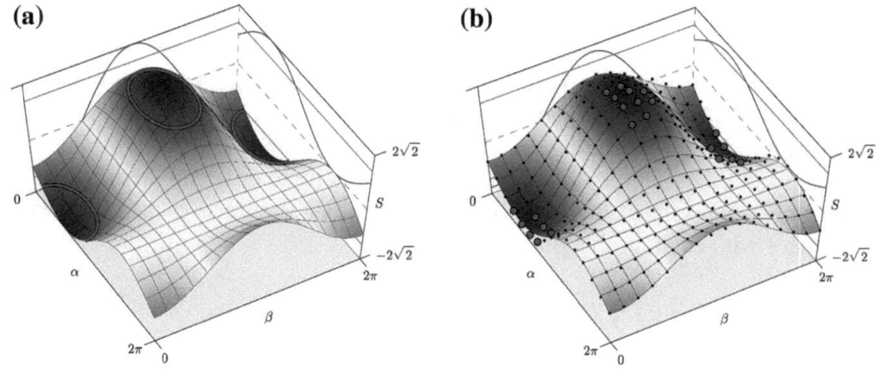

Fig. 2.14 CHSH manifold. **a** The Bell-CHSH sum S, plotted as a function of phases α and β. In the α axis, the state shared between Alice and Bob is tuned continuously between product states at $\alpha = 0, \pi$ and maximally entangled states at $\alpha = \pi/2, 3\pi/2$. The β axis shows S as a function of Bob's variable measurements, which can be thought of as two operator-axes in the real plane of the Bloch sphere, fixed with respect to each other at an angle of $\pi/2$ but otherwise free to rotate with angle β between 0 and 2π. The *blue curves* show a projection of the manifold onto each axis. Yellow contours mark the edges of regions of the manifold which violate $-2 \leq S \leq 2$. Red lines on the axes also show this limit. **b** Experimentally measured manifold. Data points are drawn as *black circles*. Data points which violate the CHSH inequality are drawn as *yellow circles*. The surface shows a fit to the experimental data (color online)

and $\beta \in [0, 2\pi]$, with step size $2\pi/15$, producing the "Bell manifold" shown in Fig. 2.14. We measured maximum and minimum values of S of 2.49 ± 0.03 and -2.54 ± 0.03 respectively. Errors were again determined by a Monte-Carlo technique, assuming Poissonian statistics.

In order to quantitatively compare the theoretical manifold with experimental data, we used the quantity

$$R^2 = 1 - \frac{\sum_i (S_i - T_i)^2}{\sum_i (S_i - \bar{S})^2}, \tag{2.57}$$

where S_i are experimentally measured values of the Bell-CHSH sum, $\bar{S}$ is the average over S_i, and T_i are the theoretical values of S shown in Fig. 2.14a. In the ideal case, $R^2 = 1$. For the data shown in Fig. 2.14b, $R^2 = 0.935$.

2.9 Generating and Characterising Mixture

Mixture, introduced in Sect. 1.3.6, is a basic property of quantum mechanical states, equivalent to classical randomness. The effect of decoherence, which is the major source of errors in many proposed architectures for quantum computing, is to introduce mixture to the computer's state, and the study and modelling of mixed states will be important in future studies of decoherence mechanisms. Despite this broad association of mixture with error, mixed states can actually be used for universal quantum computing [44], and are believed to play an important role in biological processes [45, 46] including photosynthesis.

One approach to generating mixed states is to build a source which randomly samples from an ensemble of pure states: for example, to generate the maximally-mixed single-qubit state $\mathbf{1}/2$, we can use a source which generates each of the logical basis states with equal probability

$$\hat{\rho} = \sum_i p_i |i\rangle\langle i| = \frac{1}{2}|0\rangle\langle 0| + \frac{1}{2}|1\rangle\langle 1| = \mathbf{1}/2. \tag{2.58}$$

Note that in this approach, it is important that the random sampling technique, which chooses between $|0\rangle$ and $|1\rangle$, must not "leak" information to the observer—otherwise the state can be written in a pure form:

$$|0_{t_1} 1_{t_2} 1_{t_3} 0_{t_4} 0_{t_5} 1_{t_6} \ldots\rangle \tag{2.59}$$

An alternative approach[7] begins with a maximally entangled, pure, two-qubit state, and traces over one qubit:

$$\hat{\rho}_A = \mathrm{Tr}_B \left[\frac{1}{\sqrt{2}} \left(|0_A 0_B\rangle + |1_A 1_B\rangle \right) \right] = \frac{1}{2}|0\rangle\langle 0| + \frac{1}{2}|1\rangle\langle 1| = \mathbf{1}/2. \tag{2.60}$$

The CNOT-MZ can prepare an arbitrary two-qubit state (Sect. 2.2.7), and by tracing over one qubit can thus prepare arbitrary single-qubit mixed states. Starting from the parametrization (2.25) of an arbitrary two qubit state, and tracing over the target qubit

$$|\Psi_{CT}\rangle = \sqrt{\lambda}\,|\lambda_C\rangle \otimes |\lambda_T\rangle + \sqrt{1-\lambda}\,|\lambda_C^{\perp}\rangle \otimes |\lambda_T^{\perp}\rangle \tag{2.61}$$

$$\xrightarrow{\text{trace}} \hat{\rho}_C = \mathrm{Tr}_T \left(|\Psi\rangle\langle\Psi| \right) = \lambda|\lambda_C\rangle\langle\lambda_C| + (1-\lambda)|\lambda_C^{\perp}\rangle\langle\lambda_C^{\perp}| \tag{2.62}$$

Since $|\lambda_C\rangle$ is an arbitrary single-qubit pure state, $\hat{\rho}_C$ is an arbitrary mixed state. Note that there is a one-to-one correspondence between the degree of entanglement of the initial two-qubit state and the purity of the reduced density matrix, dictated by the choice of λ.

What does it mean to "trace over the target qubit" in the context of the CNOT-MZ? Ideally, we would measure the control qubit independent of the target qubit, which in principle need not be measured at all. However, since the CNOT-P is a nondeterministic gate, we must count in the coincidence basis to postselect on successful gate operation. Therefore, in practice we count coincidences across both qubits and then

[7] Note: these two forms of mixture are sometimes distinguished as *improper* (using entangled states) and *proper* (using a random number generator). However, they are formally indistinguishable [47].

combine two-photon count-rates to generate effective single-qubit data, independent of the measurement outcome on the target:

$$\tilde{c}_{0_C} = c_{0_C 0_T} + c_{0_C 1_T}; \quad \tilde{c}_{1_C} = c_{1_C 0_T} + c_{1_C 1_T}. \tag{2.63}$$

We chose 119 single-qubit mixed states of varying purity, at random by the Hilbert-Schmidt measure [48], which samples uniformly from the full volume of the Bloch sphere. For each mixed state, we generated an appropriate two-qubit pure state, traced out the target qubit, and performed full single-qubit QST on the control, reconstructing the reduced density matrix based on $\tilde{c}_{0_C}$, $\tilde{c}_{1_C}$. Figure 2.15 shows the distribution of quantum state fidelity (2.45) between reconstructed states and their corresponding ideal mixed states. The average fidelity across all 119 states was found to be 0.98 ± 0.02, with 91 % of states having fidelity >0.95. We then chose 63 specific mixed states that mapped out the symbol 'Ψ' inside the Bloch sphere, and generated them with high fidelity (Fig. 2.15, inset). This picture gives a visual impression of the typical fidelity with which mixed states (and, by implication, entangled states) can be prepared and measured using the CNOT-MZ.

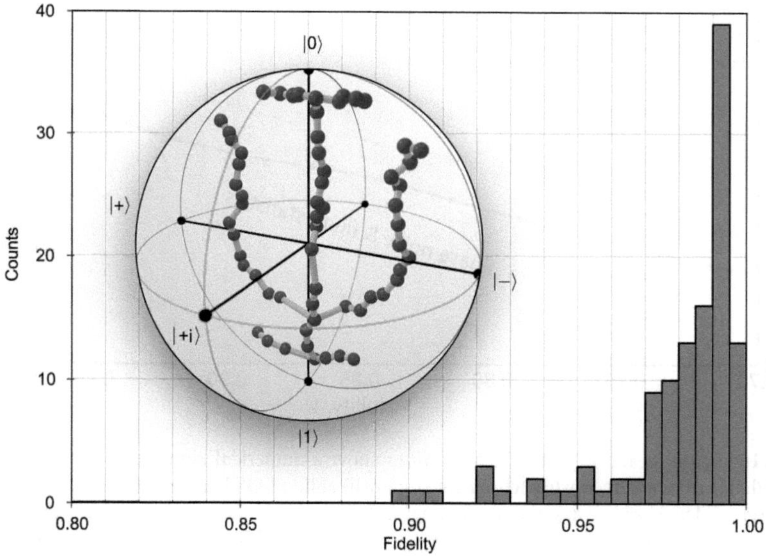

Fig. 2.15 Histogram showing the statistical distribution of quantum state fidelity between 119 randomly chosen single-qubit target states and the corresponding mixed states generated and characterized on-chip. *Inset* Ψ drawn in the Bloch sphere using 63 mixed states, again generated and characterized on-chip. These states are chosen from the real plane of the sphere for clarity. The point at the centre of the sphere is maximally mixed, and was traced out from a two-qubit maximally entangled state. Points on the surface of the sphere are pure, and were traced out from separable states

2.9.1 Errors in the CNOT-MZ

The imperfect performance of the CNOT-MZ seen in the previous experiments can be attributed to a number of different sources of error. First, we do not achieve perfect HOM interference, due to residual distinguishability of the photon pair—this is likely due to small polarization rotations, temporal distinguishability, and imperfect mode-matching at the DCs. A larger fraction of error is due to imperfect calibration and operation of the thermal phaseshifters, which contributes significantly to imperfection in reconstructed states and processes. Figure 2.16 shows the effect of inaccuracy in the control of phases in the CNOT-MZ: the fidelity of states reconstructed by QST is reduced by ~4 % given 0.05 rad of variance at each phaseshifter. We expect that imperfect fabrication of passive waveguide structures in the CNOT-MZ, which leads to time-invariant unitary errors and is reflected in the results of Sect. 2.7.1, accounts for the remaining discrepancy between our experiment and the ideal performance of the device.

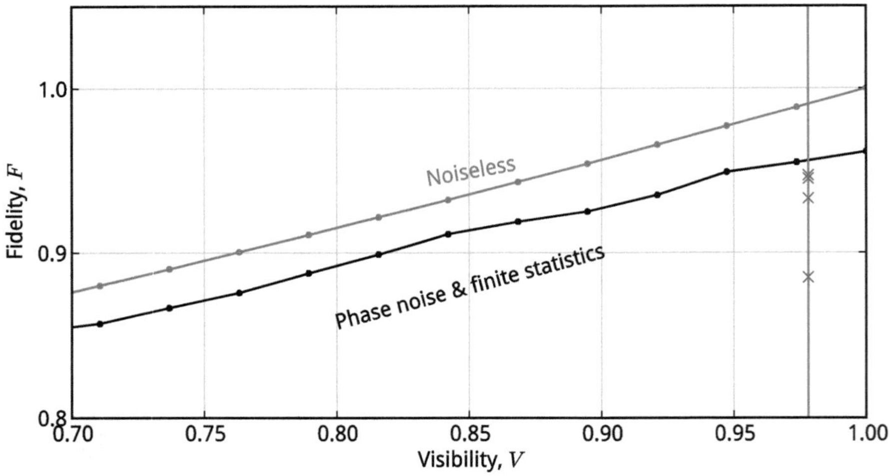

Fig. 2.16 Errors in the CNOT-MZ. *Solid lines* show a numerical simulation, plotting quantum state fidelity of states reconstructed by maximum-likelihood QST against the visibility of HOM interference. The *grey line* assumes perfect phaseshifters and infinite statistics, while the *black line* models the effect of 0.05 rad variance in phase on each phaseshifter, as well as the effects of finite statistics for a realistic experimental count-rate. The *red line* shows the experimentally measured visibility of HOM interference, and *red crosses* show measured quantum state fidelities of the four Bell states (color online)

2.10 Discussion

In this chapter, we have not shown any new ability to manipulate quantum states which could not be duplicated in practice using bulk optics. The CNOT-P gate [24], experimental state and process tomography [27], and mixed-state preparation have all previously been shown in bulk. The main result of work presented in this chapter is instead to show that the complexity and flexibility of bulk optics for quantum information can be reproduced to equivalent or better fidelity in a waveguide chip. This represents a significant step forward with respect to previous experiments in integrated quantum photonics, where devices were either completely passive [7, 9, 10, 12, 28] or insufficiently complex/reconfigurable to perform multiple distinct tasks [7, 11].

A side-effect of photonic integration is the ease with which the circuit can be fully automated, enabling experiments which depend on a large number of measurements (Chaps. 3 and 4), or feedback and optimization over a large number of experimental parameters (Chap. 5). Automation to this extent can be experimentally demanding or expensive in bulk-optics.

There remains considerable scope for improvement of the experimental setup and device fabrication. First, the silica-on-silicon material system used here is intrinsically limited by the available refractive index contrast, which leads to relatively large devices. A competitive quantum information processor built in silica-on-silicon would likely be prohibitively large. Recently, there has been great progress in integrated quantum optics using material systems which allow for a much higher component density: in particular, silicon nanowire waveguides [49–55], can provide up to six orders of magnitude decrease in component size.

As discussed in Sect. 2.9.1, inaccuracy in phaseshifter calibration is significantly detrimental to the performance of the device. Recently, Li et al. [31] have shown a new method for calibration of the CNOT-MZ, using a Bayesian learning method to automatically find the optimal calibration settings. The authors report significant improvements in the performance of the device, with respect to those reported here.

To summarize, we have shown an integrated quantum photonic chip with a considerably greater degree of reconfigurability than previous devices. We have demonstrated the ability of this chip to generate arbitrary two-qubit entangled states and single-qubit mixed states. We have confirmed the entangling capability of the device through violation of a Bell inequality across a large fraction of the parameter space. Finally, we have completely characterised the quantum process implemented by the CNOT-P gate by QPT. To our knowledge, in the field of integrated quantum photonics, this work constitutes the first demonstration of quantum state and process tomography where state preparation and measurement were both implemented on-chip, as well as the first on-chip Bell violation. The general-purpose utility of the CNOT-MZ is borne out in the following chapters.

Statement of Work

I optimized the photon source, and found and optimized the Hong-Ou-Mandel dip. I built, optimized and programmed a large fraction of the supporting electronics. I calibrated the resistive heaters, and designed and optimized the pulse sequence described in Sect. 2.3.2. I measured all of the experimental data, and performed all of the simulations shown in this section. I conceived the randomized characterization protocol.

References

1. D. Deutsch, Quantum theory, the church-turing principle and the universal quantum computer. Proc. Roy. Soc. Lond. A **400**, 97–117 (1985)
2. R. Raussendorf, H.J. Briegel, A one-way quantum computer. Phys. Rev. Lett. **86** (2001)
3. R. Raussendorf, D. Browne, H. Briegel, Measurement-based quantum computation on cluster states. Phys. Rev. A **68**(2) (2003)
4. H.J. Briegel, R. Raussendorf, Persistent entanglement in arrays of interacting particles. Phys. Rev. Lett. **86**(5), 910–913 (2001)
5. D.E. Browne, T. Rudolph, Resource-efficient linear optical quantum computation. Phys. Rev. Lett. **95**(1) (2005)
6. J. Sun, E. Timurdogan, A. Yaacobi, E.S. Hosseini, M.R. Watts, Large-scale nanophotonic phased array. Nature **493**, 195 (2013)
7. A. Politi, J.C.F. Matthews, M.G. Thompson, J.L. O'Brien, Integrated quantum photonics. IEEE J. Sel. Topics Quant. Electron. **15**, 1673–1684 (2009)
8. A. Laing, V. Scarani, J.G. Rarity, J.L. O'Brien, Reference frame independent quantum key distribution. Phys. Rev. Lett. (2010)
9. G.D. Marshall, A. Politi, J.C.F. Matthews, P. Dekker, M. Ams, M.J. Withford, J.L. O'Brien, Laser written waveguide photonic quantum circuits. Opt. Express **17**, 12546–12554 (2009)
10. A. Peruzzo, A. Laing, A. Politi, T. Rudolph, J.L. O'Brien, Multimode quantum interference of photons in multiport integrated devices. Nat. Commun. **2**, 224+ (2011)
11. J.C.F. Matthews, A. Politi, A. Stefanov, J.L. O'Brien, Manipulation of multiphoton entanglement in waveguide quantum circuits. Nat. Photon. **3**, 346–350 (2009)
12. J.C.F. Matthews, A. Politi, D. Bonneau, J.L. O'Brien, Heralding two-photon and four-photon path entanglement on a chip. Phys. Rev. Lett. **107**, 163602 (2011)
13. D. Bonneau, M. Lobino, P. Jiang, C.M. Natarajan, M.G. Tanner, R.H. Hadfield, S.N. Dorenbos, V. Zwiller, M.G. Thompson, J.L. O'Brien, Fast path and polarization manipulation of telecom wavelength single photons in lithium niobate waveguide devices. Phys. Rev. Lett. **108**(5), 053601 (2012)
14. CIP Technologies, Formerly British Telecom, now Huawei
15. D. Dai, Z. Wang, D. Liang, J.E. Bower, On-chip polarization handling for silicon photonics. SPIE Newsroom (2012)
16. A. Crespi, R. Ramponi, R. Osellame, L. Sansoni, I. Bongioanni, F. Sciarrino, G. Vallone, P. Mataloni, Integrated photonic quantum gates for polarization qubits. Nat. Commun. **2**, 566 (2011)
17. G. Lifante, *Integrated Photonics: Fundamentals* (Wiley, New York, 2003)
18. S. Lloyd, Almost any quantum logic gate is universal. Phys. Rev. Lett. **75**, 346–349 (1995)
19. D.P. DiVincenzo, Quantum computation. Science **270**(5234), 255–261 (1995)
20. T.C. Ralph, N.K. Langford, T.B. Bell, A.G. White, Linear optical controlled-NOT gate in the coincidence basis. Phys. Rev. A **65**, 062324+ (2002)

21. H.F. Hofmann, S. Takeuchi, Quantum phase gate for photonic qubits using only beam splitters and postselection. Phys. Rev. A **66**, 024308+ (2002)
22. J. Franson, B.C. Jacobs, T.B. Pittman, Quantum logic operations using linear optical elements, in *Nonlinear Optics: Materials, Fundamentals and Applications*, p. FA3 (Optical Society of America, 2002)
23. T.B. Pittman, B.C. Jacobs, J.D. Franson, Demonstration of nondeterministic quantum logic operations using linear optical elements. Phys. Rev. Lett. **88**(25), 257902 (2002)
24. J.L. O'Brien, G.J. Pryde, A.G. White, T.C. Ralph, D. Branning, Demonstration of an all-optical quantum controlled-NOT gate. Nature **426**, 264–267 (2003)
25. S. Gasparoni, J.W. Pan, P. Walther, T. Rudolph, A. Zeilinger, Realization of a photonic controlled-NOT gate sufficient for quantum computation. Phys. Rev. Lett. **93**, 020504+ (2004)
26. T.C. Ralph, Scaling of multiple postselected quantum gates in optics. Phys. Rev. A **70**, 012312 (2004)
27. J.L. O'Brien, G.J. Pryde, A. Gilchrist, D.F.V. James, N.K. Langford, T.C. Ralph, A.G. White, Quantum process tomography of a controlled-NOT gate. Phys. Rev. Lett. **93**, 080502+ (2004)
28. A. Politi, J.C.F. Matthews, J.L. O'Brien, Shor's quantum factoring algorithm on a photonic chip. Science **325**, 1221+ (2009)
29. M.A. Nielsen, I.L. Chuang, *Quantum Computation and Quantum Information (Cambridge Series on Information and the Natural Sciences)*, 1 edn (Cambridge University Press, Cambridge, 2004)
30. S.M. Tan, D.F. Walls, M.J. Collett, Nonlocality of a single photon. Phys. Rev. Lett. **66**, 252–255 (1991). Jan
31. H.W. Li, J. Wabnig, D. Bitauld, P. Shadbolt, A. Politi, A. Laing, J.L. O'Brien, A.O. Niskanen, Calibration and high fidelity measurement of a quantum photonic chip. New J. Phys. **15**(6), 063017 (2013). June
32. E. Knill, D. Leibfried, R. Reichle, J. Britton, R.B. Blakestad, J.D. Jost, C. Langer, R. Ozeri, S. Seidelin, D.J. Wineland, Randomized benchmarking of quantum gates. Phys. Rev. A **77** (2008)
33. D.F.V. James, P.G. Kwiat, W.J. Munro, A.G. White, Measurement of qubits. Phys. Rev. A **64**, 052312+ (2001)
34. G.C. Stokes, Trans. Cambr. Phil. Soc. **9**, 399 (1852)
35. M. Paris, J. Rehacek, *Quantum State Estimation* (Springer-Verlag, Berlin, 2004)
36. A. Montanaro, A. Ambainis, S. Aaronson, D. Chen, D. Gottesman, V. Liew, Three quantum learning algorithms (2013)
37. N.K. Langford, *Encoding, manipulating and measuring quantum information in optics*. Ph.D. thesis
38. W.-B. Gao, C.-Y. Lu, X.-C. Yao, P. Xu, O. Gühne, A. Goebel, Y.-A. Chen, C.-Z. Peng, Z.-B. Chen, J.-W. Pan, Experimental demonstration of a hyper-entangled ten-qubit Schrödinger cat state. arXiv:0809.4277 (2008)
39. D. Gross, Y.-K. Liu, S.T. Flammia, S. Becker, J. Eisert, Quantum state tomography via compressed sensing. Phys. Rev. Lett. **105**(15), 150401 (2010)
40. M. Cramer, M.B. Plenio, S.T. Flammia, R. Somma, D. Gross, S.D. Bartlett, O. Landon-Cardinal, D. Poulin, Y.-K. Liu, Efficient quantum state tomography. Nat. Commun. **1** (2010)
41. A.G. White, A. Gilchrist, G.J. Pryde, J.L. O'Brien, M.J. Bremner, N.K. Langford, Measuring two-qubit gates. J. Opt. Soc. Am. B **24**, 172–183 (2007)
42. G. Balló, K.M. Hangos, Parameter estimation of quantum processes using convex optimization. ArXiv e-prints (2010)
43. A. Gilchrist, N.K. Langford, M.A. Nielsen, Distance measures to compare real and ideal quantum processes. Phys. Rev. A **71**, 062310+ (2009)
44. B.P. Lanyon, M. Barbieri, M.P. Almeida, T. Jennewein, T.C. Ralph, K.J. Resch, G.J. Pryde, J.L. O/'Brien, A. Gilchrist, A.G. White, Simplifying quantum logic using higher-dimensional Hilbert spaces. Nat. Phys. **5**, 134–140 (2008)
45. M. Mohseni, P. Rebentrost, S. Lloyd, A. Guzik, Environment-assisted quantum walks in photosynthetic energy transfer. J. Chem. Phys. **129**, 174106+ (2008)

46. M.B. Plenio, S.F. Huelga, Dephasing-assisted transport: Quantum networks and biomolecules. New J. Phys. **10**(11), 113019+ (2008)
47. F. Masillo, G. Scolarici, S. Sozzo, Proper versus improper mixtures: Toward a quaternionic quantum mechanics. Theor. Math. Phys. **160**, 1006–1013 (2009)
48. K. Zyczkowski, H.-J. Sommers, Induced measures in the space of mixed quantum states. J. Phys. A. **34**, 7111+ (2001)
49. J. Silverstone, D. Bonneau, K. Ohira, N. Suzuki, H. Yoshida, N. Iizuka, M. Ezaki, R. Hadfield, G. D. Marshall, V. Zwiller, J. Rarity, J. OBrien, M. Thompson, On-chip quantum interference between two silicon waveguide sources. arXiv:1304.1490 (2013)
50. N. Matsuda, H. Le Jeannic, H. Fukuda, T. Tsuchizawa, W.J. Munro, K. Shimizu, K. Yamada, Y. Tokura, H. Takesue, A monolithically integrated polarization entangled photon pair source on a silicon chip. Sci. Rep. **2** (2012)
51. S. Azzini, D. Grassani, M.J. Strain, M. Sorel, L.G. Helt, J.E. Sipe, M. Liscidini, M. Galli, D. Bajoni, Ultra-low power generation of twin photons in a compact silicon ring resonator. Opt. Express **20**(21), 23100–23107 (2012)
52. H. Takesue, H. Fukuda, T. Tsuchizawa, T. Watanabe, K. Yamada, Y. Tokura, S. Itabashi, Generation of polarization entangledphoton pairs using silicon wirewaveguide. Opt. Express **16**(8), 5721–5727 (2008)
53. J.E. Sharping, K.F. Lee, M.A. Foster, A.C. Turner, B.S. Schmidt, M. Lipson, A.L. Gaeta, P. Kumar, Generation of correlated photons in nanoscale silicon waveguides. Opt. Express **14**(25), 12388–12393 (2006)
54. D. Bonneau, E. Engin, K. Ohira, N. Suzuki, H. Yoshida, N. Iizuka, M. Ezaki, C.M. Natarajan, M.G. Tanner, R.H. Hadfield, S.N. Dorenbos, V. Zwiller, J.L. O'Brien, M.G. Thompson, Quantum interference and manipulation of entanglement in silicon wire waveguide quantum circuits (2012)
55. A. Martin, O. Alibart, M.P. De Micheli, D.B. Ostrowsky, S. Tanzilli, A quantum relay chip based on telecommunication integrated optics technology. New J. Phys. **14**(2), 025002 (2012)

Chapter 3
A Quantum Delayed-Choice Experiment

Any other situation in quantum mechanics, it turns out, can be explained afterwards by saying, "you remember the case of the experiment with the two holes? It's the same thing."

Feyman

3.1 Introduction

This chapter concerns the fundamental concept of *wave-particle duality*. We begin with an introduction to the topic, and an overview of key results from the literature. We then demonstrate a variation on Wheeler's celebrated *delayed-choice* experiment, in which the choice of the classical observer is replaced by the state of an ancillary quantum system. This allows two mutually exclusive measurement settings to be simultaneously entertained in coherent superposition, giving rise to continuous morphing between wave-like and particle-like behaviour. Our experimental results support the understanding that the photon is neither particle nor wave, and that it does not "choose in advance" to behave as one or the other.

In this discussion I have attempted to follow closely the approach of Richard Feynman [1], and I draw on some insight due to Albert [2].

3.2 Young's Double Slit

Young's double slit is a thought experiment to do with waves and particles. By means of a simple apparatus, it reveals *the one true mystery of quantum mechanics*. Young's double slit is a "triangle" (Sect. 1.3), in the sense that it is a contrived experiment whose results cannot be elegantly explained by classical laws. Attempts are often made to shoehorn this experiment into a classical framework, but none achieve the elegance and generality of the quantum mechanical formalism. In the course of this discussion, we will see that quantum systems are neither particles nor waves, and that

P. Shadbolt, *Complexity and Control in Quantum Photonics*,
Springer Theses, DOI 10.1007/978-3-319-21518-1_3

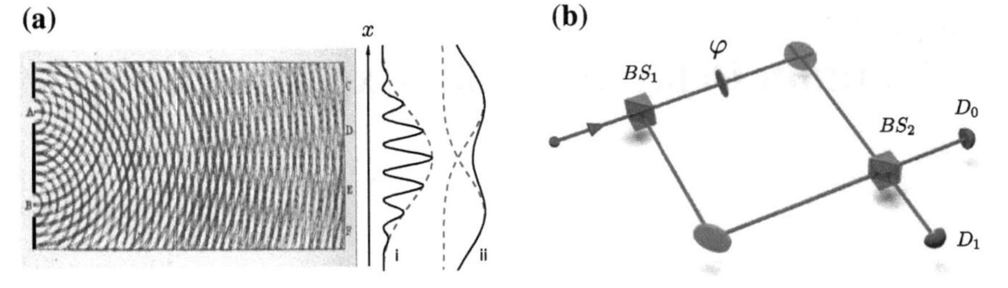

Fig. 3.1 Wave-particle duality. **a** Young's double slit experiment. Single quanta, for instance electrons or photons, are sent one-by-one towards a mask into which two holes (A, B) have been cut. On the far side of the shield the wavefunction of the particle interferes with itself, giving rise to a complex interference pattern in the distribution of detection events at an imaging screen. Wave interference drawing taken from T. Young, Course of Lectures on Natural Philosophy and the Mechanical Arts, 1807. **b** Similar interference effects are seen in the Mach-Zehnder interferometer. The intensity of light and the probability of detection, in either detector varies as a sinusoidal function of the path length difference φ in the interferometer

they are neither here, nor there, nor in two places at once, nor nowhere at all! Thus Young's double slit exposes in a very simple way the inadequacy of our everyday classical language when dealing with quantum phenomena.

Consider a machine gun, pointed at a mask in which two holes (A, B) have been made. The holes can be opened or closed at will. The gun sprays bullets across some solid angle, and from time to time a bullet will go through one or other of the holes. At a screen on the far side of the mask, a bullet detector registers the arrival of the bullet and its position on the x-axis (Fig. 3.1a). Bullets are corpuscular, indistinguishable particles, which for the purpose of this discussion are assumed to be indestructible and pass through one hole only, never both at the same time. The number of bullets arriving at the detector in a single shot is either zero or one—simultaneous detection of two bullets never occurs.

Having fired many times and detected N bullets, we can estimate the probability of detection at a particular point on the x-axis as $p_{AB}(x) = n_{AB}(x)/N$, where $n(x)$ is the total number of bullets detected at position x, and the subscript AB denotes the case where both holes are open. The full probability distribution over x consists of two overlapping lobes, corresponding to photons passing through holes A and B respectively (Fig. 3.1a, curve (ii)). If we block hole A we observe a single-lobed distribution $p_B(x)$ corresponding to photons passing through hole B only, and vice-versa. The probability distribution observed when both holes are open is equal the sum of the single-hole distributions, $p_{AB} = p_A + p_B$. This is a direct implication of the fact that bullets, being solid and lumpy, do not *interfere* with themselves.

Now we replace the machine gun with a source of waves. Perhaps stones are thrown into a lake at an appropriate distance, such that sinusoidal plane waves are incident upon the mask. These waves pass through the holes A and B, and finally arrive at a detection screen at the far side. The depth $d(t)$ of the water rises and falls continuously in peaks and troughs, and is not discrete or countable. The detection screen is sensitive only to the *average* disturbance, energy dissipation at a point, or *intensity*, a continuous variable $I_{AB}(x) \propto \int d(x,t)^2 dt$ at position x on the screen.

If we perform this experiment using water, or light, or any other kind of wave, we see a complex distribution of intensity as shown in Fig. 3.1a, curve (i). Part of this complexity is due to wave interference. A single wavefront from the source passes through *both* holes at once, giving rise to wave components originating from each of the two holes, whose peaks arrive at a given point on the screen with differing *phase* by virtue of the geometrical difference in path length. Two peaks together give a large intensity, while a peak and a trough cancel out. The function describing I_{AB} is thus composed of a $sinc$ term, corresponding to diffraction through a *single* hole, and a sinusoidal term due to wave interference between the two holes. If we block hole B, $I_A(x)$ reduces to a $sinc$ function only, and all interference effects disappear. As a result, $I_{AB} \neq I_A + I_B$, in strong contrast with bullets.

What happens if we repeat this experiment using a source of quantum particles? Here we will discuss photons, but essentially identical results are observed for electrons, atoms, and even large molecules such as C_{60} (buckminsterfullerene) [3]. We take a source which, upon pressing a button, generates a single photon—the Fock state $|1\rangle = a^\dagger|0\rangle$. A single-photon detector (see Sect. 1.6.4) is arranged at a position x on the far side of the mask, in the plane of the screen. A photon is sent towards the holes, and with some probability $p_{AB}(x)$ the detector will *click*, generating an electrical pulse. This output is binary—either the detector clicks, absorbing $\hbar\omega$ of energy, or it does not. Using a true single-photon (Fock-state) source, simultaneous detection of a photon at two separate detectors is never observed (see Sect. 1.5.2). In this sense, photons behave very much like particles. They arrive at the detector as corpuscular, indivisible lumps, and it is natural to think that they might also *travel* as such, physically passing through one hole or the other.

Having fired many photons and registered N detection events, as with bullets, we begin to saturate the probability distribution $p_{AB}(x) = n_{AB}(x)/N$. If photons are entirely particle-like, we expect to see two lobes, as in curve (ii). Instead, we measure probability distributions with the exact form of curve (i)! If we block one or other of the holes, we recover single-lobed $sinc$-like probability distributions, as with water waves. Thus the photonic probability distribution does *not* obey $p_{AB} = p_A + p_B$, and can only be described in terms of wave interference between components arising simultaneously from holes A and B. Now we encounter a serious philosophical problem.

It is natural to ask: *where was the photon* when it passed through the holes? Did it travel through a single hole, as a particle, or both, as a wave? If we take two detectors and place them inside holes A and B, we only ever detect the photon at one hole or the other, never registering a detection event in both holes simultaneously. This must be true for energy to be conserved. Now,

- If the photon passed through *one hole only*, and did not pass through the other, we cannot explain the wave interference effects observed.

- If the photon passed through *both holes simultaneously*, as if it were a wave, then it stands to reason that we could detect it at both holes simultaneously, which *never* occurs.
- If the photon does not pass through either hole, then we would never detect it at all—but we do.

So, the photon does not pass through hole A nor hole B alone, and it does not pass through both holes simultaneously, and it does not pass through *neither* hole—but it nevertheless arrives at the screen! In this experiment, the photon exhibits *wave-particle duality*, seemingly travelling and arriving as a lump, as if it were a particle, but simultaneously exhibiting wave interference—phenomena which are classically mutually exclusive. Contained in this experiment is the full mystery of quantum mechanics.

An optical implementation of Young's double slit experiment was performed in 1909 by Sir Geoffrey Taylor, who used a gas flame together with smoked glass[1] to generate "feeble light", and observed interference fringes in the shadow cast by a sewing needle [4]. In 1961, the experiment was first performed using electrons [5]. More recent experimental results include double-slit interference of Buckminster-fullerene [3], and an electron interference experiment using micromachined slits [6] which could be opened and closed at will.

3.2.1 Wave-Particle Duality in the MZI

Throughout the rest of this chapter, it will be convenient to modify the experimental arrangement somewhat with respect to Feynman's original proposal. Figure 3.1b shows a Mach-Zehnder interferometer (MZI, Sect. 1.5.4), which exhibits all of the essential behaviour of Young's double slit, but is somewhat easier to analyse. Single photons are sent into one input port of BS_1, pass through the two paths of the interferometer and interfere with themselves at BS_2. The two arms of the interferometer have a path length difference φ. BS_1 assumes the role of the shield and holes, and BS_2 provides an interface at which the two beams may interfere, in a similar role to the screen. Two detectors, D_0 and D_1, record single-photon detection events at each output port of BS_2. The probability of detecting a photon at a given detector is a sinusoidal function of φ:

$$p(D_0) = \cos^2\left(\frac{\varphi}{2}\right) \ ; \ p(D_1) = \sin^2\left(\frac{\varphi}{2}\right). \tag{3.1}$$

In this interference pattern we clearly see the wavelike properties of the photon.

[1] The intensity of light in Taylor's experiment was roughly equivalent to a candle burning at a distance of one mile. J.J. Thompson's expectation, which turned out to be incorrect, was that the diffraction pattern should be modified in the limit of very low light levels, as the corpuscular nature of the photon appeared.

In the double-slit scenario discussed previously, the screen is deliberately placed at a considerable distance from the shield such that diffraction patterns from the two slits overlap at the screen. Detection of a photon at a point x therefore does not yield any information about which path (hole) was taken. It is easy to see that if the screen is placed in the near-field, without any overlap, full which-way information is obtained upon detection—but no interference (wave-like) effects are seen. An analogous choice of measurement setting can be performed in the MZI. When BS_2 is removed from the interferometer, every detection event tells the observer whether the photon took the upper or lower path—full which-way information—but the interference pattern necessarily cannot be observed. In this case the detection probabilities no longer depend on φ:

$$p(D_0) = \frac{1}{2}\,; \quad p(D_1) = \frac{1}{2} \tag{3.2}$$

If we define this mode of operation as fully particle-like behaviour, then we can view the removal of BS_2 as switching from wave-like to particle-like measurement apparatus, where each configuration reveals a complementary aspect of the photon.

3.2.2 Complementarity

Niels Bohr's *complementarity* is the fundamental physical principle at the heart of the Copenhagen interpretation of quantum theory, and enforces limitations on the interface between quantum systems and the classical data available to an experimentalist. It states that in order to observe complementary properties of a quantum system, an experimentalist must necessarily employ mutually incompatible arrangements of the measurement apparatus. Complementarity was characterized by Bohr as follows:

> "…it is only the mutual exclusion of any two experimental procedures, permitting the unambiguous definition of complementary physical quantities, which provides room for new physical laws" [7]

In Young's double slit, as we have already seen, we can arrange the apparatus so as to measure particle-like behaviour of the photon, *watching* it take one path or the other. However, in order to see wavelike interference effects from which the phase φ can be inferred, we must adopt an experimentally incompatible measurement setup, obscuring all which-way information. That we cannot use both measurement setups at once is not merely a consequence of inadequate apparatus, or lack of imagination on behalf of the experimentalist. It is simply a consequence of the fact that experimental data is by definition classical—pencil marks on a piece of paper, or magnetic domains on a hard disk—and cannot therefore exist in quantum superposition. Thus, as was emphasized by Bohr, a single configuration of any given measurement apparatus may only reveal *part* of the quantum mechanical phenomenon. Only by use of multiple configurations or instances of the classical measurement apparatus is the fullness of wave-particle duality, or any other quantum effect, revealed.

Bohr's principle has only very recently been successfully quantified in *universal complementarity relations*, such as those due to Ozawa and Hall [8, 9]. It was shown that if two incompatible observables $\hat{A}$ and $\hat{B}$, $[\hat{A}, \hat{B}] \neq 0$ are approximated by $\hat{A}_{est}$ and $\hat{B}_{est}$, $[\hat{A}_{est}, \hat{B}_{est}] = 0$, then the rms error $\epsilon(\hat{G}_{est}) \equiv \langle(\hat{G}_{est} - \hat{G})^2\rangle^{1/2}$ in measurements of these observables must satisfy

$$\epsilon(\hat{A}_{est})\epsilon(\hat{B}_{est}) + \epsilon(\hat{A}_{est})\Delta\hat{B} + \Delta\hat{A}\epsilon(\hat{B}_{est}) \geq \frac{c}{2} \tag{3.3}$$

where $\epsilon(\hat{G}_{est}) \equiv (\langle\hat{G}^2\rangle - \langle\hat{G}\rangle^2)^{1/2}$ is the *spread* in the quantity G. This formalizes the notion that although the inaccuracy in either observable can individually be made arbitrarily small, one cannot simultaneously measure both to an arbitrary degree of accuracy. This relation was recently experimentally tested by Weston et al. [10] under conditions in which previously discovered, non-universal complementarity relations fail.

Complementarity lies at the heart of the Copenhagen interpretation of quantum mechanics. In contrast with de Broglie-Bohm carrier-wave theory [11] (in which the photon has a literal particle-like trajectory even when unobserved) and the many-worlds interpretation due to Everett [12] (in which wavefunction collapse does not occur at all), complementarity states that the photon is neither particle-like nor wave-like until it is measured, at which point the wavefunction collapses in accordance with the choice of measurement apparatus.

3.3 Wheeler's Delayed Choice Experiment

Upon first encountering Young's double slit experiment, many physicists are disturbed by its implications. This discomfort does not typically reduce as a function of time—with greater understanding it should increase! It is nonetheless natural to attempt to find comfort in a classical understanding of the experiment, where meaningful comparison can be drawn between the behaviour of the photon and that of everyday objects in the macroscopic world.

One such classical explanation is very simple to imagine, if somewhat extravagant in conception. Let us allow that the photon is *sentient*, or is otherwise able to examine and assess the experimental apparatus prior to measurement. If the photon determines that the measurement device is arranged so as to reveal particle-like behaviour—that is, BS_2 is removed from the interferometer—then before it reaches BS_1, the decision is made to become *fully particle-like*, throwing away all wave-like properties. Upon arrival at BS_1 the photon chooses one path or the other, exactly as though it were a particle. It then propagates through the apparatus with impunity, ultimately reproducing exact particle-like statistics: $p(D_0) = p(D_1) = 1/2$. If BS_2 is instead present, corresponding to a wave-like measurement, the photon decides in advance to adopt a fully wave-like nature. Wave interference is then observed at the

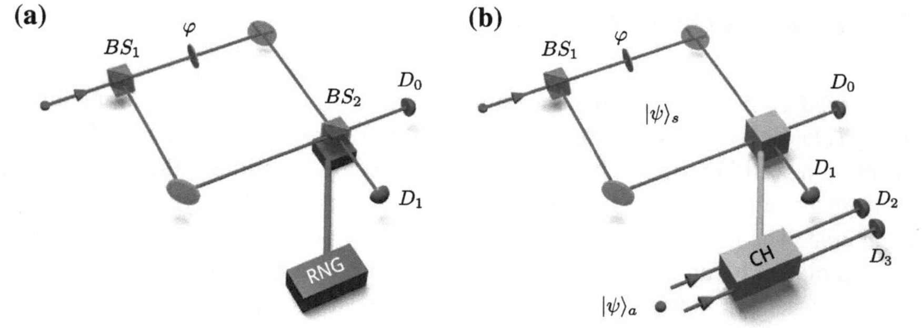

Fig. 3.2 **a** Wheeler's delayed choice experiment. A photon is sent into a Mach-Zehnder interferometer. Upon arrival at the first beamsplitter BS_1, it is split into quantum superposition across both paths. A space-like separated random number generator (*RNG*) then toggles a fast optical switch, closing or opening the interferometer by insertion or removal of BS_2, leading to wave-like or particle-like measurement of the photon respectively. Two detectors, D_0 and D_1, reveal wave-like behaviour in the event that the interferometer is closed, otherwise particle-like statistics are seen. **b** Quantum delayed choice. The optical switch is replaced by a quantum-controlled beamsplitter: a controlled-Hadamard gate. An ancilla photon controls this gate: ancilla states $|0\rangle$ and $|1\rangle$ lead to presence and absence of BS_2 respectively. By preparing the ancilla in a superposition state, BS_2 is effectively placed into a superposition of present and absent, leading to a superposition of wave-like and particle-like measurement

detectors, from whose output the phase φ may be inferred, without any need for the photon to choose a particular path upon arrival at BS_1.

Complementarity and the necessity of incompatible measurement devices make it difficult to distinguish this pseudo-classical hidden-variable model from the quantum mechanical reality. A particularly elegant approach, which makes life very hard for the sentient photon, was proposed by Wheeler in 1978 [13, 14]. The trick in Wheeler's *delayed-choice* experiment, shown in Fig. 3.2a, is to postpone the choice of measurement apparatus until such time as the photon is inside the interferometer. Once the photon has passed BS_1, a fast classical switch is used to remove or insert BS_2 at will. Now, upon arrival at BS_1, the photon must choose to behave as particle or wave *without prior knowledge of the measurement apparatus*. Hence, if it is true that the photon adopts the pathological classical behaviour described above, we expect to see a deviation from the quantum predictions.

Delayed-choice experiments have been performed in a variety of physical systems [15–19], all of which confirm the quantum predictions. Of particular significance is a recent result [19] of Jacques et al., in which relativistic space-like separation between the random choice of measurement setting and the entry point of the interferometer (BS_1) was achieved for the first time. Here, a nitrogen vacancy colour centre in diamond was used as the source of single photons, ensuring extremely close approximation to the Fock state $|1\rangle$. An electro-optic phaseshifter, controlled by a quantum random number generator at 4.2 MHz, was used to implement the choice of measurement setting.

3.4 Quantum Delayed Choice

In delayed-choice experiments, the choice of the observer is generally implemented using a classical optical switch, fast enough to effectively insert or remove BS_2 while the photon is still in flight. This classical-controlled beamsplitter is driven by a single bit from a random number generator, or the free and independent choice of the experimentalist. The main distinguishing feature in our work is that the classical random bit is replaced by an ancilla qubit $|\psi\rangle_a$, which drives a *quantum-controlled beamsplitter*, as shown in Fig. 3.2b. This configuration was first proposed in a theoretical work due to Ionicioiu and Terno [20],

It is helpful in this analysis to note that Wheeler's interferometer and photon together form a path-encoded qubit (see Sect. 1.6.1), where the $|1\rangle_s$ and $|0\rangle_s$ states correspond to a photon in the upper and lower arms of the interferometer respectively. In our experiment the ancilla qubit is also path-encoded, and is implemented using a second photon. We will refer to these as the *system* and *ancilla* photon/qubit/interferometer respectively.

Upon arrival at BS_1, the system photon splits into a coherent superposition over the upper and lower spatial modes of Wheeler's interferometer,

$$|\psi\rangle_s = \hat{BS_1}|0\rangle_s = \frac{1}{\sqrt{2}}\left(|0\rangle_s + |1\rangle_s\right), \tag{3.4}$$

and is then phase-shifted due to the path-length difference φ

$$|\psi\rangle_s \xrightarrow{\varphi} |\psi_{particle}\rangle_s = \frac{1}{\sqrt{2}}(|0\rangle_s + e^{i\varphi}|1\rangle_s). \tag{3.5}$$

If the ancilla qubit is prepared in the state $|0\rangle_a$, the quantum-controlled beamsplitter does not act, and BS_2 is effectively absent. The interferometer is thus left open, and the final state of the system is simply given by (3.5). In this case, the probability of detecting the system photon in either detector is $p(D_0) = p(D_1) = |\langle 0|\psi\rangle_s|^2 = 1/2$. Every detection event yields full which-way information, and no wave interference is observed.

If the ancilla is instead prepared in $|1\rangle$ the quantum-controlled beamsplitter always acts on the system qubit, closing the interferometer. This gives rise to wave interference such that

$$|\psi\rangle_s \xrightarrow{BS_2} |\psi_{wave}\rangle_s = \cos\frac{\varphi}{2}|0\rangle_s + \sin\frac{\varphi}{2}|1\rangle_s. \tag{3.6}$$

The probability that the system photon is detected at D_0 is now a sinusoidal function of the phase, $p(D_0) = \cos^2\left(\frac{\varphi}{2}\right)$, and $p(D_1) = \sin^2\left(\frac{\varphi}{2}\right)$. Formally, the quantum-controlled beamsplitter is then equivalent to the controlled-Hadamard operation CH— a maximally-entangling two-qubit gate—acting on the system qubit, with the ancilla as the control:

$$U_{CH} = |0_a0_s\rangle\langle 0_a0_s| + |0_a1_s\rangle\langle 0_a1_s| + |1_a+_s\rangle\langle 1_a0_s| + |1_a-_s\rangle\langle 1_a1_s|$$
$$= \begin{pmatrix} 1 & 0 & 0 & 0 \\ 0 & 1 & 0 & 0 \\ 0 & 0 & \frac{1}{\sqrt{2}} & \frac{1}{\sqrt{2}} \\ 0 & 0 & \frac{1}{\sqrt{2}} & \frac{-1}{\sqrt{2}} \end{pmatrix}. \tag{3.7}$$

When the ancilla qubit is prepared in a generalized superposition state

$$|\psi\rangle_a = \cos(\alpha)|0\rangle_a + \sin\alpha|1\rangle_a, \tag{3.8}$$

the second beamsplitter BS_2 is effectively placed in a coherent superposition of present and absent. The global state of the two qubits then evolves to

$$|\Psi_f(\alpha,\varphi)\rangle = \cos\alpha\ |0\rangle_a \otimes |\psi_{particle}(\varphi)\rangle_s + \sin\alpha\ |1\rangle_a \otimes |\psi_{wave}(\varphi)\rangle_s \tag{3.9}$$

which is entangled for $0 < \alpha < \pi/2$—maximally so for $\alpha = \pi/4$ and $\phi = \pi/2$. The detection probability at D_0 is given by

$$p(D_0)(\varphi,\alpha) = p(D_0)_{particle}\cos^2\alpha + p(D_1)_{wave}\sin^2\alpha$$
$$= \frac{1}{2}\cos^2\alpha + \cos^2(\frac{\varphi}{2})\sin^2\alpha \tag{3.10}$$

and $p(D_1) = 1 - p(D_0)$. Hence, in contrast with traditional implementations of Wheeler's delayed choice experiment, we are able to tune coherently and *continuously* between particle-like ($\alpha = 0$) and wave-like ($\alpha = \pi$) statistics.

An important distinguishing feature of the quantum delayed-choice setup concerns the ordering of events. Note that since the dynamic classical switch of Wheeler's traditional experiment is replaced by a static controlled-unitary operation, there is no longer any "delayed choice" in this delayed-choice experiment, and there is no need for fast switching. If the ancilla is (for example) prepared in an equal superposition, it travels balistically through the device without any explicit choice of measurement setting ever being made. Before either photon is detected, the choice of measurement setting remains in coherent superposition, encoded in the *entangled* state of system and ancilla. Only when the ancilla is detected does the wavefunction collapse to one or other measurement setting. As a result, the specific timing of the choice measurement setting is inconsequential, and can even be performed *after* the system photon has been detected.

3.4.1 Experimental Setup

As we have already noted, the quantum delayed-choice arrangement of [20] can be seen as a system of two path-encoded qubits, where the quantum-controlled beam-splitter is implemented by a CH gate. It turns out that the CNOT-MZ device described in Chap. 2 provides all the necessary hardware and a sufficient degree of reconfigurability to implement the experiment, as outlined in Fig. 3.3.

As in Chap. 2, two photons from an SPDC source are used to encode two qubits in pairs of waveguides. Wheeler's interferometer is implemented using the state preparation stage of the target qubit. The system photon is coupled into the chip, whereupon it is split across two paths by dc_2, a 50/50 directional coupler. Wheeler's phase, φ, is controlled by a thermal phaseshifter (ϕ_2). The ancilla photon is injected into the upper two waveguides on the device (previously referred to as the control qubit). State preparation of the $|\psi\rangle_a$ is accomplished using the MZI formed by directional couplers dc_1 and dc_3, where phase shifter ϕ_1 controls the α parameter.

The CH gate is implemented using the non-deterministic postselected linear-optical CZ gate previously described (couplers 6, 7, 8, Sect. 2.2.4). The CH gate is equivalent to CZ up to local rotations. Specifically,

$$U_{CH} = (I \otimes W)U_{CZ}(I \otimes W) \tag{3.11}$$

where

$$W = \begin{pmatrix} \cos\frac{\pi}{8} & \sin\frac{\pi}{8} \\ \sin\frac{\pi}{8} & -\cos\frac{\pi}{8} \end{pmatrix}. \tag{3.12}$$

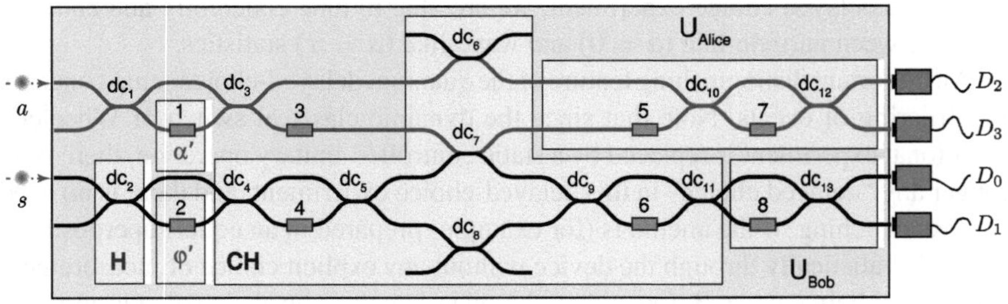

Fig. 3.3 The CNOT-MZ provides all the necessary hardware and a sufficient degree of reconfigurability to implement a quantum delayed-choice experiment. Wheeler's interferometer is mapped to the target qubit of the CNOT-MZ, with dc_2 and ϕ_2 in the roles of BS_1 and the internal phase shift φ respectively. The ancilla qubit is prepared using dc_1 and dc_3 together with phase shifter ϕ_1. The quantum-controlled beamsplitter is constructed from the linear optical CZ gate—three directional couplers (dc_6, dc_7 and dc_8) with coupling ratio $1/3$—and single-qubit W gates implemented using dc_4, dc_5, dc_9, and dc_{11} together with ϕ_4 and ϕ_6. For certain values of α and φ, the output state of the CH gate is entangled. By measuring each qubit in a particular set of measurement bases controlled using U_{Alice} and U_{Bob}, we are able to violate a Bell inequality on $|\Psi_f\rangle_{as}$, thus ruling out local hidden variable models in which the photon decides in advance to behave as a particle or wave

Implementing W gates using directional couplers dc_4, dc_5, dc_9 and dc_{11} together with phaseshifters ϕ_4 and ϕ_6, we can thus implement a controlled-Hadamard gate on the system qubit, effectively closing or opening the interferometer containing ϕ_2 depending on the state of the ancilla. As with linear-optical CZ and CNOT-P gates, this gates succeeds with $1/9$ probability and its operation depends on high visibility quantum interference, requiring that the ancilla and system photons are indistinguishable in all degrees of freedom.

Four silicon APDs are used to detect single photons at the output of the chip. As before, we only register a subset of two-photon coincidence events (D_0D_2, D_0D_3, D_1D_2, D_1D_3) so as to post-select on successful operation of the entangling gate.

3.4.2 Results

When sweeping the phase φ in Wheeler's interferometer, we should see qualitatively different behaviour of the system photon depending on the ancilla phase α. Specifically, when $\alpha = \pi/2$ we expect to see a sinusoidal wave interference pattern in the probability of detection at D_0 and D_1, while for $\alpha = 0$ we should see no interference. For intermediate values of α, we expect continuous morphing between wave-like and particle-like behaviour, as the effective probability amplitude for the presence of BS_2 is gradually reduced. Experimental data exhibiting this effect is shown in Fig. 3.4. We measured $p(D_0)$ and $p(D_1)$ for 21 values of φ in the interval $[\pi/2, 5\pi/2]$ and 11 values of α in the interval $[0, \pi/2]$.

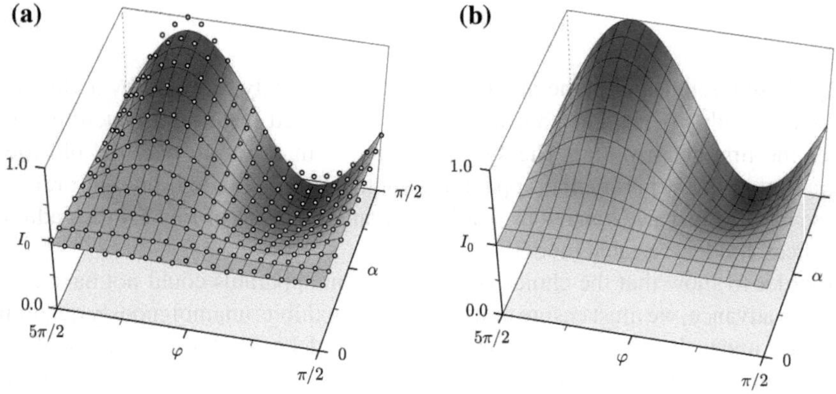

Fig. 3.4 Continuous morphing between wave-like and particle-like behaviour of the system photon, as a function of the state of the ancilla qubit $|\psi(\alpha)\rangle_a$. **a** Experimentally measured probability of detection at D_0, conditional on detection of a second photon at either D_2 or D_3 (*white dots*). The surface is a fit to the data, using Eq. 3.10 with an additional prefactor to account for limited visibility of quantum interference. **b** Ideal (simulated) behaviour

A similar experiment was carried out at the same time [21] by Kaiser et al. in the group of Sébastien Tanzilli (Nice). In contrast with our work, the authors make use of polarization entanglement directly from the SPDC source, rather than implementing a non-deterministic linear-optical entangling gate, and do not make use of path-encoding. The authors measure morphing between particle and wave behaviour qualitatively identical to the result shown in Fig. 3.4, again with extremely good agreement between experiment and theory. The decision to use polarization encoding in this implementation is largely motivated by the fact that stable Mach-Zehnder interferometers are difficult to construct in a bulk architecture. This perhaps highlights the fact that the technological advances of integrated quantum photonics, although intended primarily as a route to scalable quantum computation and practical quantum technologies, also provide advantages for more fundamental scientific investigations.

3.5 Device-Independent Tests of Wave-Particle Duality

The principal goal of delayed-choice experiments is to test the classical, hidden-variable hypothesis that the photon decides in advance to behave as a particle or a wave. Although as experimentalists we place a certain amount of trust in the notion that the behaviour of our experimental apparatus is repeatable and consistent, we must concede that the result shown in Fig. 3.4 does not absolutely rule out the hidden-variable model. Even though we have good reason to believe that the ancilla qubit is truly placed in the coherent superposition (3.8), the morphing behaviour in Fig. 3.4 could also be explained if it is instead prepared in the mixed state

$$\hat{\rho}_a = \cos^2(\alpha)|0\rangle\langle 0|_a + \sin^2(\alpha)|1\rangle\langle 1|_a. \tag{3.13}$$

Under these circumstances the ancilla qubit can be equally replaced by a classical random bit with $p(0) = \cos^2(\alpha)$, whose state is decided *before* the system photon passes the first beamsplitter. The system photon is thus free to play the old trick of examining the experimental apparatus—including this random bit—in order to choose particle or wave behaviour in advance, and the result of Fig. 3.4 thus admits a classical, hidden-variable model.

In order to show that the choice of measurement apparatus could not have been known in advance, we must ensure that the CH gate exhibits unambiguously quantum behaviour under the circumstances of the quantum delayed-choice experiment. As we have already seen, the output of the CH gate is ideally pure and entangled for almost all values of φ and α. As a result, we can test for quantum behaviour in a device independent way—that is, without having to place any trust in the measuring apparatus—by attempting to violate a Bell inequality (Sect. 1.3.8) using the bipartite state of the system and ancilla photon.

3.5.1 Results

Experimentally, we give the ancilla qubit to Alice, who chooses from one of two measurement bases using the interferometer U_{Alice} formed by dc_{10} and dc_{12}, together with phaseshifters ϕ_5 and ϕ_7 and detectors D_2 and D_3. The system photon is assigned to Bob, who performs local measurements using U_{Bob}: dc_{13} together with ϕ_8 and detectors D_0 and D_1. The choice of measurement operators $\hat{A}_{0,1}$, $\hat{B}_{0,1}$ was tailored for the specific class of states generated in the quantum delayed-choice scenario—the operators usually chosen for Bell-CHSH with the singlet state do not lead to violation here. We measured the Bell-CHSH parameter $S(\varphi, \alpha)$ over the same parameter space used in Fig. 3.4, measuring a maximal violation $S(\pi/2, \pi/4) = 2.45 \pm 0.03$. Experimental data is shown together with a simulation in Fig. 3.5.

Had we been able to perform the Bell test without succumbing to any loopholes, we could now conclude decisively that the photon does not choose in advance to behave as a particle or a wave. However, a loophole-free Bell inequality remains experimentally out of reach—although progress continues to be made [22, 23]—and our experiment does not in fact close *any* of the standard loopholes. For instance, we make the standard fair-sampling assumption, which allows us to discard inconclusive results and post-select on successful operation of the CH gate. The detection loophole remains open due to limited detection efficiency, and we must also assume independence between the operation of the photon source and the choice of measurement setting used in the Bell inequality test. As usual, if the photons could know in advance the choice of measurement setting in the Bell test, then a local model can mimic Bell inequality violations.

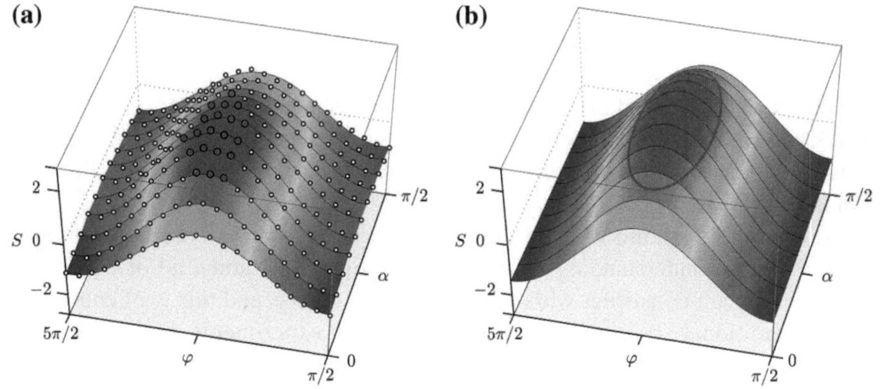

Fig. 3.5 CHSH parameter S as a function of the phase φ in Wheeler's interferometer and the ancilla parameter α. All local hidden variable models satisfy $|S| \leq 2$. **a** Experimental data (*white points*), with a 2D sinusoidal fit. Points marked in yellow exhibit nonlocal statistics, violating Bell-CHSH. **b** Numerical simulation of ideal behaviour

3.5.2 Discussion

The Greek philosopher Democritus (c. 460 BC)—proponent of atomistic theory, scourge of Plato, and staunch advocate of cheerfulness—is quoted by Schrödinger as having said, with respect to the fundamental makeup of the universe,

> "By convention there is sweetness, by convention bitterness, by convention colour, in reality only atoms and the void." [24]

Democritus goes on to emphasize the importance of *measurement*, and the difficulty with which experimental results are reconciled with our internal understanding of the world:

> "Foolish intellect! Do you seek to overthrow [the senses], while it is from [them] that you take your evidence?"

Even earlier, Lucretius (c. 99 BC) assigned a particle-like character to light:

> "The light and heat of the sun; these are composed of minute atoms which, when they are shoved off, lose no time in shooting right across the interspace of air in the direction imparted by the shove."

The history of science has since been marked by intense debate between particle and wave theories of physics, in particular with respect to the nature of light. In *Opticks* [25], Isaac Newton describes a great many experiments exploring the "reflections, refractions, inflections and colours of light". Despite the emphasis of this work on optical wave phenomena, a central hypothesis is the corpuscular nature of light, in whose defence Newton cites the tendency to travel in straight lines and cast stark shadows—"light does not bend into the shadow". This understanding was later contested by the wave theories of Huygens, Young, and Maxwell in particular, whose theory of electromagnetic waves proved so powerful as to render the corpuscular theory untenable. In the first decade of the 20th century, new explanations for the troublesome behaviour of of black-body radiation and the photoelectric effect, due to Planck and Einstein respectively, gave new legs to the idea of an indivisible particle of light with energy $\hbar\omega$, the photon, and ultimately lead to the quantum theory of light used throughout this thesis.

So, does the quantum delayed-choice experiment described here add anything to our scientific understanding of the nature of light? Certainly, all of our experimental results are consistent with known quantum theory, and this is of course true for the "traditional" delayed-choice and double-slit experiments. Since we do not close all possible loopholes, our Bell-CHSH inspired test does not achieve device-independence, although it certainly strengthens the argument that the CH gate functions as advertised. It would be interesting to perform a more refined version of our experiment, with space-like separation of Alice and Bob and with loopholes closed, although it seems unlikely that this will be technologically feasible very soon. I think that it is important to ask whether the quantum delayed-choice setup teaches us anything about the photon over and above that which can be inferred from photonic Bell-CHSH tests. Can we construct self-consistent theories of quantum mechanics, in which the photon decides in advance to behave as a particle or a wave (in some meaningful sense), but which nonetheless permit Bell-CHSH violation? If not, then my impression is that this work provides a useful and attractive pedagogical tool, but nothing more.

Statement of work

All of the experimental data presented here was measured jointly by Alberto Peruzzo and myself.

References

1. R. Feynman, R. Leighton, M. Sands, The Feynman Lectures on Physics: Volume 1 (1963)
2. D.Z. Albert, *Quantum Mechanics and Experience* (Harvard University Press, Cambridge, 1994)
3. M. Arndt, O. Nairz, J. Vos-Andreae, C. Keller, G. van der Zouw, A. Zeilinger, Wave-particle duality of c60 molecules. Nature **401**(6754), 680–682 (1999)
4. G.I. Taylor, Interference fringes with feeble light. Proc. Camb. Philos. Soc. **15**, 114–115 (1909)
5. C. Jönsson, Electron diffraction at multiple slits. Am. J. Phys. **42**, 4–11 (1974)
6. R. Bach, D. Pope, S.-H. Liou, H. Batelaan, Controlled double-slit electron diffraction. New J. Phys. **15**(3), 033018 (2013)
7. N. Bohr, Can quantum-mechanical description of physical reality be considered complete? Phys. Rev. **48**, 696–702 (1935). October
8. M.J.W. Hall, Prior information: how to circumvent the standard joint-measurement uncertainty relation. Phys. Rev. A **69**, 052113 (2004)
9. M. Ozawa, Universally valid reformulation of the heisenberg uncertainty principle on noise and disturbance in measurement. Phys. Rev. A **67**, 042105 (2003). April
10. M.M. Weston, M.J.W. Hall, M.S. Palsson, H.M. Wiseman, G.J. Pryde, Experimental test of universal complementarity relations. Phys. Rev. Lett. **110**, 220402 (2013)
11. D. Bohm, B.J. Hiley, *The Undivided Universe: An Ontological Interpretation of Quantum Theory* (Routledge, London, 1995)
12. H. Everett, Theory of the Universal Wavefunction. Ph.D. thesis, Princeton University, 1956
13. J.A. Wheeler, *Mathematical Foundations of Quantum Mechanics* (Academic, New York, 1978)
14. J.A. Wheeler, *Quantum Theory and Measurement* (Princeton University Press, New Jersey, 1984)
15. C.O. Alley, O.G. Jacubowicz, W.C. Wickes, in *Proceedings of the Second International Symposium on the Foundations of Quantum Mechanics*, ed. by H. Narani (Physics Society of Japan, Tokyo, 1987)
16. T. Hellmuth, H. Walther, A. Zajonc, W. Schleich, Delayed-choice experiments in quantum interference. Phys. Rev. A **35**, 2532–2541 (1987)
17. B.J. Lawson-Daku, R. Asimov, O. Gorceix, Ch. Miniatura, J. Robert, J. Baudon, Delayed choices in atom stern-gerlach interferometry. Phys. Rev. A **54**, 5042–5047 (1996)
18. Y.-H. Kim, Y. Rong, S.P. Kulik, Y. Shih, M.O. Scully, Delayed "choice" quantum eraser. Phys. Rev. Lett. **84**, 1–5 (2000)
19. V. Jacques, E. Wu, F. Grosshans, F. Treussart, P. Grangier, A. Aspect, J.-F. Roch. Experimental realization of wheeler's delayed-choice gedanken experiment. Science **315**(5814), 966–968 (2007)

20. R. Ionicioiu, D.R. Terno, Proposal for a quantum delayed-choice experiment. Phys. Rev. Lett. **107**, 230406 (2011). Dec
21. F. Kaiser, T. Coudreau, P. Milman, D.B. Ostrowsky, S. Tanzilli, Entanglement-enabled delayed-choice experiment. Science **338**(6107), 637–640 (2012)
22. T. Scheidl, R. Ursin, J. Kofler, S. Ramelow, X. Ma, T. Herbst, L. Ratschbacher, A. Fedrizzi, N.K. Langford, T. Jennewein, A. Zeilinger, Violation of local realism with freedom of choice. P. Natl. Acad. Sci. U.S.A **107**(46), 19708–19713 (2010)
23. M. Giustina, A. Mech, S. Ramelow, B. Wittmann, J. Kofler, J. Beyer, A. Lita, B. Calkins, T. Gerrits, S.W. Nam, R. Ursin, A. Zeilinger, Bell violation using entangled photons without the fair-sampling assumption. Nature **497**, 227–230 (2013). May
24. W. Durant, *The Story of Civilization: Part II: The Life of Greece* (Simon and Schuster, New York, 1939)
25. I. Newton, *Opticks, or a Treatise of the Reflections, Refractions, Inflections and Colours of Light*. (William Innys, 1704)

Chapter 4
Entanglement and Nonlocality Without a Shared Frame

4.1 Introduction

In many quantum information tasks, the basic scenario is one of two parties, Alice and Bob, who share an entangled state $|\psi_{AB}\rangle$ originating from a source. Alice and Bob may wish to use this state to communicate securely (Sect. 1.4.2), violate a Bell inequality (Sect. 1.3.8), perform teleportation, tomography (Sect. 2.6), or to evaluate the degree of entanglement of the state (Sect. 1.3.7). Perhaps they are space-like separated, maybe they are in the same lab, perhaps $|\psi_{AB}\rangle$ is a resource state in a quantum computer—we have already discussed many such scenarios.

One assumption that is very often made in theoretical works is that Alice and Bob share a reference frame. That is, they agree on a coordinate system in which Bob's "up" is the same as Alice's, and they can, for example, measure qubits in the $\hat{\sigma}_{x,y,z}$ bases. This assumption is often valid—in proof-of-principle experiments we usually operate within the frame of the laboratory, and have classical tools at our disposal to precisely calibrate and align Alice and Bob with respect to one another. However, there are many real-world scenarios in which full calibration and alignment is not possible.

In single-mode optical fiber, natural and unavoidable fluctuations in temperature and stress give rise to unknown, random, unitary rotations of the polarization of transmitted light [1]. These rotations roughly span the entire space of $SU(2)$—although not in any uniform way—and largely preclude the use of polarization encoding in classical telecommunications. Optical satellite links, proposed as a real-world target for photonic quantum communication [2–4], suffer from continuous rotation of the satellite with respect to earth, as well as timing drift, necessitating complex tracking and correction systems (Fig. 4.1). Path encoded qubits in bulk suffer from thermal/acoustic phase instability, which gives rise to unknown random unitary rotations of the qubit reference frame. Even if the setup is perfectly stable, we sometimes just do not have the time or tools to calibrate phaseshifters, waveplates, and polarization controllers. As quantum technologies become increasingly complex, these issues will not disappear.

P. Shadbolt, *Complexity and Control in Quantum Photonics*,
Springer Theses, DOI 10.1007/978-3-319-21518-1_4

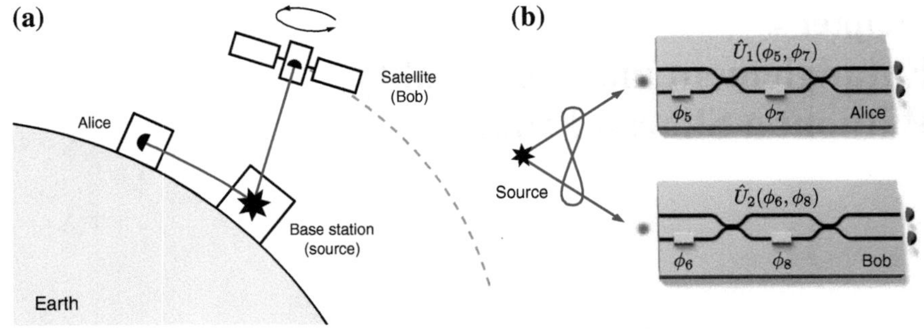

Fig. 4.1 Bell violations with random measurements. **a** A source generates entangled pairs, and photons are sent to Alice and Bob respectively. We consider a scenario in which Alice and Bob do not share a frame of reference—that is, they cannot choose a common measurement basis—and are therefore forced to measure in randomly oriented bases. **b** We use the CNOT-MZ to experimentally test a scheme which guarantees Bell inequality violation even in the absence of a shared reference frame. Path-entangled photon pairs are generated by the CNOT-P gate and measured in a qubit basis by Alice and Bob, who are implemented using the readout stage of the CNOT-MZ. The choice of measurement setting is accomplished using thermal phase shifters ϕ_{5-8}

In all such scenarios, unknown rotations decouple Alice's reference frame from Bob, effectively *breaking* most tomographic protocols, entanglement witnesses, Bell tests, and QKD. Sometimes we can use active stabilization or further classical communication to establish a shared frame, but it is interesting to ask—how well can we perform these QI tasks in the absence of any shared reference frame?

In this chapter we show how detection of Bell nonlocality can be guaranteed—preserving device independence—without a shared frame, even in the absence of well-calibrated devices. We experimentally demonstrate that by randomizing voltages on the CNOT-MZ, we can violate a Bell inequality with high probability. Finally, we describe a practical method to accurately measure the degree of entanglement of a two-qubit state despite time-dependent unitary noise on the local channel between source and observer. This method makes direct use of Haar-random noise to improve performance, and allows an experimentalist to detect entanglement by simply shaking, bending and twisting non-polarization maintaining optical fiber. We discuss possible applications of this scheme to measurement and secure communication.

4.2 Bell Tests Without a Shared Frame

In Sects. 1.3.8, 1.4.2, and 2.8 of this thesis we have seen the significance of nonlocality as a fundamental quantum mechanical phenomenon, as well the utility of nonlocal correlations as a tool for device-independent quantum communication and state characterization. Bell tests such as Bell-CHSH, described in detail in Sect. 1.3.8, provide an experimental prescription for rigorous certification of nonlocal statistics.

It will be convenient to first re-write the Bell-CHSH inequality (1.42) using a slightly different notation. We assume that Alice and Bob measure a two-qubit state $|\psi\rangle$ using m local measurement settings per party, $\hat{A}_j, \hat{B}_j, i, j \in [0, m-1]$ respectively. For $m = 2$, all local hidden variable (LHV) models must satisfy the Bell-CHSH inequality

$$|S| = |\langle \hat{A}_0 \hat{B}_0 \rangle + \langle \hat{A}_0 \hat{B}_1 \rangle + \langle \hat{A}_1 \hat{B}_0 \rangle - \langle \hat{A}_1 \hat{B}_1 \rangle| \leq 2. \tag{4.1}$$

Since the indexing of each measurement setting $\hat{A}_i, \hat{B}_j$ is arbitrary, terms in (1.42) can be possibly permuted, moving the minus sign and creating a number of equally valid Bell inequalities. Local-realistic models satisfy all such permutations. For instance, $|\langle \hat{A}_1 \hat{B}_0 \rangle + \langle \hat{A}_1 \hat{B}_1 \rangle + \langle \hat{A}_0 \hat{B}_0 \rangle - \langle \hat{A}_0 \hat{B}_1 \rangle| \leq 2$ holds for all LHV theories. Violation of any of the 36 allowed inequalities witnesses nonlocal behaviour.

For two qubits, although entanglement is necessary in order to obtain nonlocal statistics, it is not sufficient.[1] Even if Alice and Bob share a maximally entangled state, they will not necessarily violate CHSH if they do not *measure* in appropriate bases. To see this, first assume that Alice and Bob share the maximally entangled Bell state $|\Psi^-\rangle$. Letting $\hat{A}_0 = \hat{\sigma}_x, \ \hat{A}_1 = \hat{\sigma}_z, \ \hat{B}_0 = \hat{\sigma}_x, \ \hat{B}_1 = \hat{\sigma}_z$, it is simple to show that $S = 0$, yielding no violation, and no nonlocal correlations. However, if we rotate $\hat{B}_j$,

$$\hat{B}_0 = \frac{\hat{\sigma}_x + \hat{\sigma}_z}{\sqrt{2}}, \quad \hat{B}_1 = \frac{\hat{\sigma}_x - \hat{\sigma}_z}{\sqrt{2}} \tag{4.2}$$

we recover maximal violation of CHSH, $|S| = 2\sqrt{2}$. In the theoretical discussion of such scenarios it is often implicitly assumed that Alice and Bob share a reference frame. How does CHSH perform when there is no common frame?

4.2.1 Theory

Let us assume that Alice and Bob share the singlet state $|\Psi^-\rangle$, but have no information that would allow them to establish a shared reference frame, and that they are interested in violating Bell-CHSH with the greatest possible efficiency. In this discussion it will be useful to consider the measurement settings of Alice and Bob in terms of their Bloch vectors (1.47) $\vec{a}_i, \ \vec{b}_j \in \mathbb{R}^3$. Alice and Bob each choose two vectors $\vec{a}_{0,1}$ and $\vec{b}_{0,1}$, independently from a uniform distribution over the 2-sphere (equivalent to the Haar measure for $SU(2)$, see Sect. 1.3.1), and measure in all combinations of $\hat{A}_i \hat{B}_j$. In 2010, Liang et al. showed [6] that Bell violation is achieved in this scenario with a probability of $\sim$28 %. If Alice and Bob are each able to choose *mutually unbiased* vectors, orthonormal in the Bloch sphere and obeying $\vec{a}_i \cdot \vec{a}_j = \delta_{ij}$, this probability increases to $\sim$42 %. This analysis has been generalized to the multipartite case [6, 7], as well as to schemes involving decoherence-free subspaces [8]. These

[1] Note that the picture is more complex for multi-particle scenarios, where nonlocality can be seen without entanglement. See for example [5].

results show that it is more probable to detect nonlocality than one might naïvely expect. However, can it be guaranteed?

So far we have allowed only two measurement settings per party. Consider now a scenario in which each party chooses *three* settings, $\hat{A}_{0,1,2}$ and $\hat{B}_{0,1,2}$ where we again demand that these measurements are mutually unbiased, thus forming randomly-oriented orthogonal triads in the Bloch spheres of Alice and Bob respectively. It turns out that in this situation, we can *always* find a valid Bell inequality of the form (4.1) which is violated. In other words, by adding one measurement setting per party, detection of nonlocality can be guaranteed.

Proof Assume that $\vec{a}_{i\in\{0,1,2\}}$ and $\vec{b}_{i\in\{0,1,2\}}$ are mutually unbiased vectors corresponding to qubit measurement operators $\hat{A}_i$ and $\hat{B}_j$. Alice and Bob evaluate these expectation values for the singlet state $|\Psi^-\rangle$ over all combinations of i, j and can then write them in matrix form,

$$\mathcal{E} = \begin{pmatrix} E_{00} & E_{01} & E_{02} \\ E_{10} & E_{11} & E_{12} \\ E_{20} & E_{21} & E_{22} \end{pmatrix}. \tag{4.3}$$

It is straightforward to show that these expectation values are given by the scalar product $E_{ij} = \langle \hat{A}_i \hat{B}_j \rangle = -\vec{a}_i \cdot \vec{b}_j$. The columns of $\mathcal{E}$ are therefore equivalent to the coordinates b'_k of the Bloch vectors $\vec{b}_j$ in the basis $\vec{a}_i$. By re-labelling measurement settings and outcomes we are free to permute rows and/or columns of this matrix as well as possibly change their sign. We can therefore assume, without loss of generality, that $E_{00,11,22} > 0$, and that E_{22} is the largest element by absolute value in $\mathcal{E}$. $\vec{b}_j$ are orthonormal, giving $\vec{b}_2 = \pm\vec{b}_0 \times \vec{b}_1$ and therefore $|E_{22}| = |E_{00}E_{11} - E_{01}E_{10}|$. Now, $E_{22} = E_{00}E_{11} - E_{01}E_{10} \geq E_{00}, E_{11}, |E_{01}|, |E_{10}|$ and $E_{01}E_{10} \leq 0$. We assume that $E_{01} \leq 0$ and $E_{10} \geq 0$, if this is not the case then we are free to multiply the second row and column by -1. Now we have that

$$(E_{00} + E_{10}) \max[-E_{01}, E_{11}] \geq E_{00}E_{11} - E_{01}E_{10} = E_{22} \geq \max[-E_{01}, E_{11}]. \tag{4.4}$$

Dividing by $\max[-E_{01}, E_{11}] > 0$, we find $E_{00} + E_{10} \geq 0$. Using a similar method, we can show $-E_{01} + E_{11} \geq 0$. Adding these inequalities, we obtain

$$E_{00} + E_{10} - E_{01} + E_{11} \geq 1\,. \tag{4.5}$$

By construction, $\mathcal{E}$ is an orthogonal matrix. Therefore, this inequality is satisfied if and only if $E_{00} + E_{10} = 0$, $-E_{01} + E_{11} = 0$ and $\vec{a}_0 = \vec{b}_0$, $\vec{a}_1 = \vec{b}_1$ and $\vec{a}_2 = \vec{b}_2$. That is, so long as Alice's measurements are not *perfectly* aligned with respect to Bob's, CHSH is violated. □

An independent proof of this result was obtained by Wallman and Bartlett [9], and published shortly after our manuscript appeared in *Scientific Reports*.

We have shown that CHSH can be violated with certainty without a shared reference frame, when Alice and Bob share a perfect maximally entangled state. However,

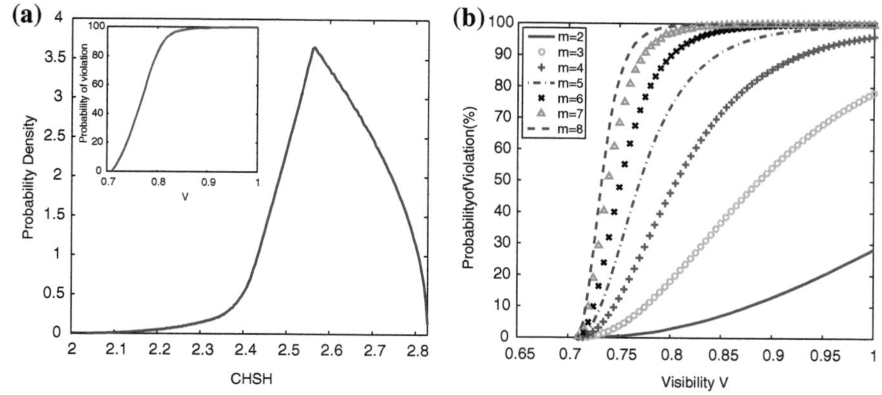

Fig. 4.2 **a** Bell tests using random measurement triads. Numerically computed distribution of maximum CHSH violation for uniformly random, mutually unbiased measurement triads on a singlet state. **b** Bell tests using completely random measurements, without calibration. Numerical calculation of the probability of Bell violation as a function of Werner state visibility V, for different numbers m of uniformly random measurements per party

in order for this scheme to be practically relevant, we must consider its performance under realistic experimental imperfections.

Experimental Bell tests are necessarily limited to measuring finite statistics, resulting in uncertainty in measured expectation values. This gives rise to error in S, and we must therefore examine the distribution of CHSH over all allowed measurement settings to ensure that the probability of violation remains high despite such uncertainty. Figure 4.2a shows a numerical calculation of the distribution of $|S|$ when $\vec{a}_i$, $\vec{b}_j$ are constructed around a random vector chosen by the Haar measure. The distribution is perhaps surprisingly weighted towards large violation, with a mean value $\bar{S} \sim 2.6$. In order to take into account experimental uncertainty δ in S we can shift the local bound $\mathcal{L}$, modifying Bell-CHSH as

$$|S| \leq \mathcal{L} = 2 + \delta. \tag{4.6}$$

Even with $\delta = 0.2$, corresponding to only a few hundred detection events, the probability of violation for a perfect singlet state remains at ~99.7 %.

Of course, entangled states in prepared in the lab are never perfect. We use a partially mixed Werner state (1.51), whose purity is characterised by the *visibility* V, to model this imperfection. Note that this is not equivalent to the visibility of quantum interference (1.134). Figure 4.2(a, inset) shows the probability of violation as a function of V, demonstrating the robustness of generic nonlocality to imperfect experimental state preparation. For example, with $V = 0.9$ and $\delta = 0.1$, the probability of violation remains greater than 98.2 %.

4.2.2 Experiment

The scheme described here is immediately applicable to a broad variety of scenarios, physical systems, and qubit encodings, including polarization states of entangled photons in optical fiber and free space, and path-encoding in photonic chips. We chose to perform our experimental implementation using the CNOT-MZ chip previously described, providing two path-encoded qubits with arbitrary state preparation and measurement capabilities. The scheme for reference-frame independent Bell violation described here is not absolutely necessary in order to violate CHSH on the CNOT-MZ, as alignment of reference frames is relatively straightforward. However, as we show in the Sect. 4.3, an extension to this scheme allows Bell violation with the CNOT-MZ in a "black-box" scenario, using completely uncalibrated phaseshifters.

We experimentally tested the situation in which Alice and Bob measure the singlet using orthogonal measurement triads. We prepare the singlet state using indistinguishable photons from a type-I SPDC source, together with the CNOT' gate and local rotations as described in Sect. 2.6.3. We then generate randomly chosen measurement triads $\vec{a}_i, \vec{b}_j$ using a pseudo-random number generator [11]. Having calibrated the phase/voltage relationship of the phase shifters as described in Sect. 2.3.3, we then apply appropriate voltages to phaseshifters ϕ_{5-8} in order to perform the nine measurements, evaluating E_{ij}. For each measurement setting, two-photon coincidence counts between all 4 combinations of APDs (C_{00}, C_{01}, C_{10}, C_{11}) are then measured for a fixed amount of time. The typical rate of simultaneous photon detection coincidences was ~1 kHz. From this data we compute the maximal CHSH value as detailed above, and the entire procedure is repeated 100 times. The results are presented in Fig. 4.3a, where accidental coincidences, arising primarily from photons originating from different down-conversion events, which are measured throughout the experiment, have been subtracted from the data. Remarkably, all 100 trials lead to a clear CHSH violation; the average CHSH value we observe is ~2.45, while the smallest measured value is ~2.10.

The visibility of the highest-fidelity experimental state was 0.913 ± 0.004, measured by maximum-likelihood quantum state tomography. Experimental imperfection in the photon source, CNOT-MZ device, and phaseshifter calibration all account for reduced visibility of the state, as described in Sect. 2.9.1. In order to further test the robustness of the reference-frame-independent scheme described here, we deliberately introduced a temporal delay between the two photons at the SPDC source, increasing their distinguishability. The effect is as though the CNOT' gate implements an incoherent mixture of the CNOT' and identity operations [12]. Note that this does not reproduce the Werner state $\hat{\rho}_V$, instead approximating $\hat{\rho} = p|\Psi^-\rangle\langle\Psi^-| + (1-p)|01\rangle\langle 01|$, where p depends non-trivially on the temporal delay.

We repeat the protocol described above for a range of visibilities, estimating the visibility of the state through tomographic reconstruction of the experimental density matrix. Figure 4.3b clearly demonstrates the robustness of our scheme, in good agreement with theoretical predictions: a considerable amount of mixture must be

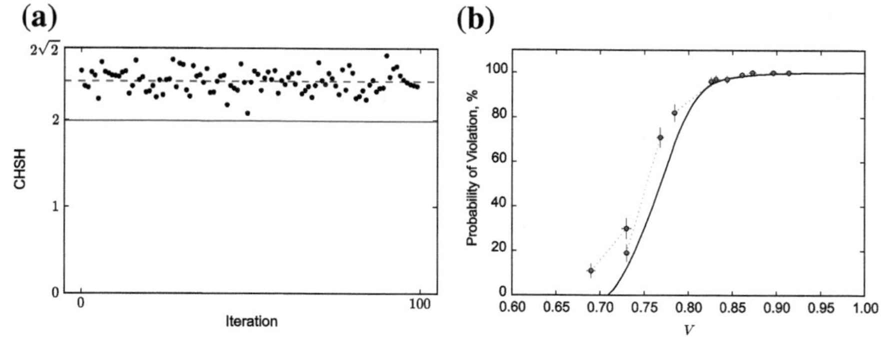

Fig. 4.3 Bell tests requiring no shared reference frame. **a** 100 successive Bell tests. In each iteration, both Alice and Bob use a randomly-chosen measurement triad. For each iteration, the maximal CHSH value is plotted (*black points*). We observe CHSH violation in all trials; the *red line* indicates the local bound ($S = 2$). The smallest CHSH value is ∼2.1, while the mean CHSH value (*dashed line*) is ∼2.45. This leads to an estimate of the visibility of $V = \frac{2.45}{2.6} \simeq 0.942$, to be compared with 0.913 ± 0.004 obtained by maximum likelihood quantum state tomography [10]. This slight discrepancy is due to the fact that our entangled state is not exactly of the form of a Werner state. Error bars are too small to draw. **b** The experiment of (**a**) is repeated with reduced visibility of quantum interference, illustrating the robustness of the scheme. Each point shows the probability of CHSH violation estimated using 100 trials. Uncertainty in probability is estimated as the standard error. Visibility for each point is estimated by state tomography, where the error bar is calculated using a Monte Carlo approach. *Red points* show data corrected for accidental coincidences. The *black line* shows the theoretical curve from Fig. 4.2

introduced in order to significantly reduce the probability of obtaining a CHSH violation. The discrepancy between experiment and theory is largely due to tomographic errors and the fact that we do not exactly prepare the Werner state (1.51). Together these results show that large Bell violations can be obtained without a shared reference frame, even with realistic experimental imperfections.

4.3 Bell Tests Without Calibrated Devices

Although Alice and Bob do not need to share a reference frame in order to implement the scheme described above, they nonetheless require *well-calibrated* measurement devices in order to construct mutually unbiased measurement triads. Calibration of measurement devices, such as wave-plates, phaseshifters, etc. is a routine task, but may be challenging or even impossible in certain scenarios, forcing Alice and Bob to measure in completely random, non-orthogonal bases, which are unlikely to be uniformly distributed on the 2-sphere.

It was shown in [6] that if Alice and Bob choose measurements entirely at random, the probability of violation is $p \sim 28\,\%$. If they make n repeated measurements of S using random settings, they will asymptotically approach unit probability of eventual violation as $P_n \sim 0.72^n$. Can they do better than this?

4.3.1 Theory

Assume that Alice and Bob measure all possible expectation values E_{ij} over m random measurement settings per party $\vec{a}_i$ and $\vec{b}_j$. We can again write these expectation values in matrix form,

$$\mathcal{E} = \begin{pmatrix} E_{00} & E_{01} & E_{02} & E_{03} & \dots \\ E_{10} & E_{11} & E_{12} & E_{13} & \dots \\ E_{20} & E_{21} & E_{22} & E_{23} & \dots \\ E_{30} & E_{31} & E_{32} & E_{33} & \dots \\ & & & & \ddots \\ \dots & \dots & \dots & \dots & \end{pmatrix} \tag{4.7}$$

Now, there are an enormous number of ways in which groups of four expectation values from $\mathcal{E}$ can be combined to form valid CHSH inequalities in the form of (4.1). As a result, although it is no longer guaranteed that we will obtain nonlocal statistics, the probability of violation increases rapidly with m to the extent that for $m = 5$, assuming a perfect singlet state, $p \sim 99.5\,\%$. This approach is similarly robust to limited visibility of the state, yielding $\sim$97 % probability of violation when $V = 0.9$ and $m = 5$. Figure 4.2 shows the results of numerical simulations of this scenario for $m \in [2, 8]$.

4.3.2 Experiment

Although the phaseshifters in the CNOT-MZ device had been well calibrated prior to this experiment, we emphasise the time-consuming nature of the calibration procedure, and the fact that the phase-voltage relationship is not consistent across heaters. To further complicate calibration, the phase-voltage response of an individual heater will drift with use over time. In order to demonstrate the robustness of the above scheme to non-uniform randomness in the choice of measurement settings, we performed Bell-CHSH tests *without* making use of the available phase-voltage information for phaseshifters ϕ_{5-8}.

Having prepared the singlet state using the CNOT' gate, we chose the measurement operators $\vec{a}_i$ and $\vec{b}_j$ by randomly picking voltages in the interval $[0, 7]$ V for phaseshifters $\phi_{5,6}$ and $\phi_{7,8}$ respectively, where 7 V is simply a hardware limitation of the heaters. Since the phase-voltage response of each heater is nonlinear (see Sect. 2.3.3), this gives rise to phases which are not uniformly distributed in the interval $[0, 2\pi]$, and therefore measurement bases $\hat{A}_i$, $\hat{B}_j$ which are certainly not chosen by the Haar measure.

We implemented this protocol for $m \in \{2, 3, 4, 5\}$, observing a rapid increase in the probability of violation with m, as shown in Fig. 4.4. For $m = 5$, we find 95 out of 100 trials lead to a CHSH violation, even when the choice of measurement is not uniformly random. The visibility V of the state used for this experiment was measured using state tomography to be 0.869 ± 0.003.

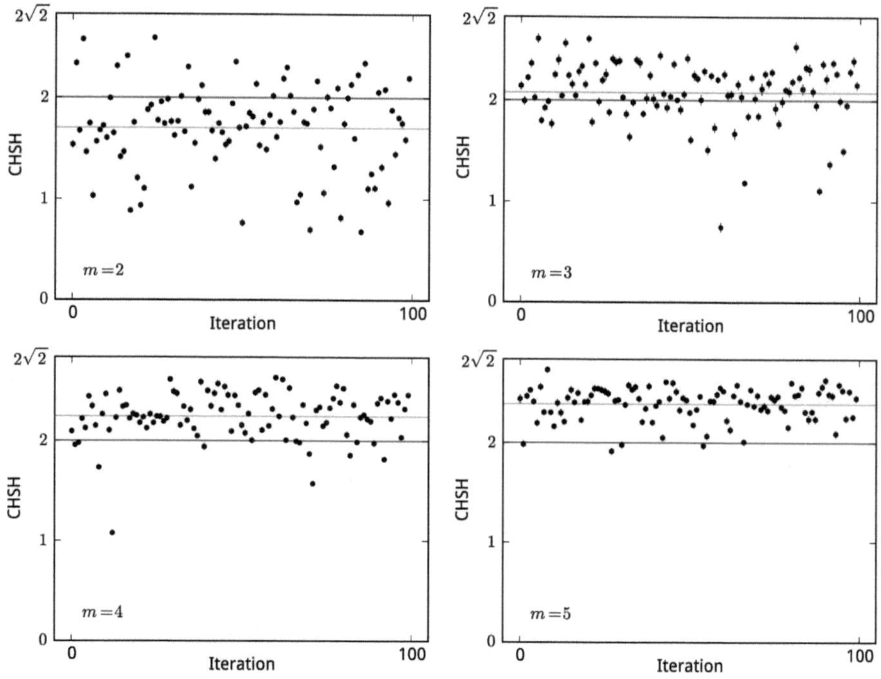

Fig. 4.4 Experimental Bell tests using uncalibrated devices. We perform Bell tests on a two-qubit Bell state using uncalibrated measurement interferometers, choosing voltages uniformly from the interval $[0, 7]\,V$. For $m = 2, 3, 4, 5$ local measurement settings, we perform 100 trials (for each value of m). As the number of measurement settings m increases, the probability of obtaining a Bell violation rapidly approaches one. For $m \geq 3$, the average CHSH value (*dashed line*) is above the local bound of CHSH$=2$ (*red line*). Error bars were estimated by a Monte Carlo technique, assuming Poissonian statistics. This data has been corrected for accidental coincidences (color online)

4.4 Discussion

Often, entanglement and nonlocality are seen as rare and fragile phenomena, extremely sensitive to experimental noise and imperfection. By showing that nonlocality can be robustly detected without the need to calibrate or align measurement devices, even with limited visibility of state preparation, we have provided a new fundamental insight into the generic nature of nonlocality.

The schemes described here potentially have practical applications. First, Bell tests provide an unambiguous and device-independent test for the presence of entanglement—a powerful tool for the future development of quantum technologies—and the ability to perform such tests without calibration or alignment will likely facilitate such tests in some scenarios. The necessary criteria for a loophole-free reference-frame independent Bell test using the scheme described in Sect. 4.2 are discussed in further detail by Gómez et al. [13], paying particular attention to detection efficiencies. A further experimental implementation of Bell tests using orthogonal triads has been performed by Palsson et al. [14]. It has recently been proposed [15] that this scheme might also facilitate the detection of nonlocality of a *single* photon (see, for example, Ref. [16]).

We also expect that this work will find applications in quantum communication protocols. Previous work by Laing et al. [17] describes a technique for reference-frame independent QKD in which the parties share in advance a single common measurement basis, and this theory has recently been implemented in collaboration with Nokia [4, 18]. Our work extends this capability to the case where no knowledge of the reference-frame is shared. More recent interest in the general topic of alignment-free quantum communication has been reviewed by D'Ambrosio et al. [19], and an experimental implementation of device-independent QKD without a shared reference frame, using our theoretical framework, was recently described in [20].

4.5 A Noise-Powered Entanglement Detector

In the previous discussion, although the relative orientation of Alice and Bob's frames is unknown, it was assumed that this orientation does not change between consecutive measurements, i.e. every element of (4.3) can be estimated before $\vec{a}_i, \vec{b}_j$ change by any appreciable amount. However, for realistic unstable reference frames—including polarization in fiber and bulk path interferometers—the rate of change is often so high that this assumption does not hold.

We now consider the situation in which each reference frame changes, uniformly and at random, every time Alice and Bob measure an expectation value. Naïvely, it might appear that there is then very little that Alice and Bob can say about the state, as they are forced to make observations through a "fog" of random, uncorrelated local unitary rotations. However, we will introduce a simple protocol which *exploits* this noise, allowing Alice and Bob to distinguish between entangled and separable sources, and estimate certain physical properties of the state. We discuss immediate practical applications of this scheme with respect to state characterization and secure communication.

The experimental scenario is illustrated in Fig. 4.5. Charlie has a source, which generates qubit pairs in the state $\hat{\rho}_C$. He claims that $\hat{\rho}_C$ is entangled, but this claim is not trusted. Charlie sends qubit pairs to two observers, Alice and Bob, who measure their respective systems in a local basis before comparing measurement outcomes. The channels between Charlie and Alice/Bob, corresponding to unitary operators $\hat{U}_A(t)$, $\hat{U}_B(t)$, are assumed to be lossless but unstable. At some time t, the two-qubit channel $\hat{\mathcal{U}}$ is described by a unitary operator

$$\hat{\mathcal{U}}(t) = \hat{U}_A(t) \otimes \hat{U}_B(t), \tag{4.8}$$

where $\hat{U}_A(t)$, $\hat{U}_B(t)$ are chosen independently and at random from the Haar measure on $SU(2)$. After some interval Δt, instability in the channel leads to new instances of

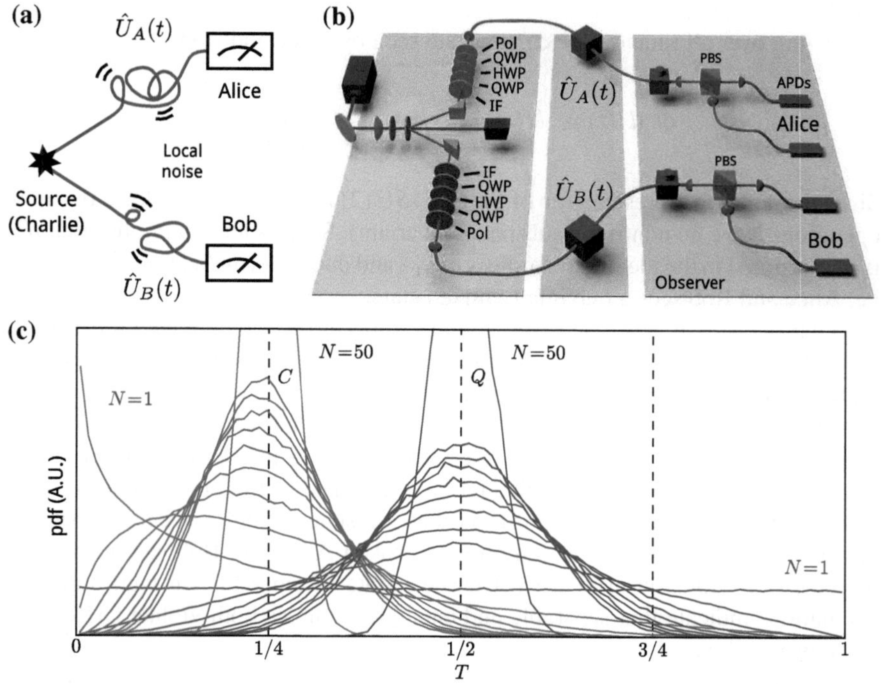

Fig. 4.5 **a** Charlie has an untrusted source of two-qubit states $\hat{\rho}_C$, which he claims is entangled. Alice and Bob want to reliably estimate some measure of entanglement generated at the source, but their view is obscured by an unstable unitary channel. **b** Photonic experimental implementation. Charlie has a type-2 SPDC source of photon pairs, which can be switched between entangled and separable operation. He sends photon pairs to Alice and Bob through non-polarization-maintaining optical fiber, which is continuously moved, bent, and twisted throughout the experiment. **c** Numerical simulation, showing the distribution of $T = \frac{1}{N}\sum_i |\langle ZZ\rangle_i|$, when Charlie prepares any separable state (*red lines*) versus any maximally entangled state (*blue lines*). Alice and Bob are thus able to distinguish entangled and separable sources

$\hat{U}_A, \hat{U}_B$, drawn again from the Haar measure. During a single timestep $t_j = t_0 + j\Delta t$, Alice and Bob receive n copies of the state

$$\hat{\rho}^i_{AB} = \hat{\mathcal{U}}(t_j)\ \hat{\rho}_C\ \hat{\mathcal{U}}(t_j)^\dagger = \hat{\mathcal{U}}_j\ \hat{\rho}_C\ \hat{\mathcal{U}}_j^\dagger \tag{4.9}$$

where n is sufficiently large to give a good estimate of the expectation value

$$E_j = \langle \hat{A}_j \otimes \hat{B}_j \rangle = \mathrm{Tr}\,(\hat{\rho}^j_{AB}\ \hat{A}_j \otimes \hat{B}_j), \tag{4.10}$$

where $\hat{A}$, $\hat{B}$ are Alice and Bob's single-qubit measurement operators respectively.

Alice and Bob would now like to determine whether Charlie's state is entangled. Note that Charlie is not held accountable for the behaviour of the channel—Alice and Bob care about the degree of entanglement of $\hat{\rho}_C$, which is independent of $\hat{\mathcal{U}}$. Averaging over all time ($t \to \infty$), the state seen by Alice and Bob is

$$\hat{\rho}_{AB}^{\infty} = \int_0^{\infty} \mathrm{d}t \, (\hat{\mathcal{U}}(t) \, \hat{\rho}_C \, \hat{\mathcal{U}}(t)^{\dagger}) = \int_{SU(2) \otimes SU(2)} \mathrm{d}\hat{\mathcal{U}} \, (\hat{\mathcal{U}} \, \hat{\rho}_C \, \hat{\mathcal{U}}^{\dagger}). \tag{4.11}$$

Since the defining representation of $SU(2) \otimes SU(2)$ is irreducible (all local two qubit operations leave no nontrivial subspaces invariant), Schur's lemma implies that $\hat{\rho}_{AB}^{\infty}$ is proportional to the identity regardless of $\hat{\rho}_C$, and due to normalization, $\hat{\rho}_{AB}^{\infty} = \mathbf{1}/4$, i.e. Alice and Bob sees a maximally mixed state.

What happens if Alice and Bob attempt to perform quantum state tomography (Sect. 2.6), ignoring fluctuations in the channel? Since tomography depends on a finite number of measured expectation values ($t < \infty$), the reconstructed density matrix is not necessarily maximally mixed, but nevertheless provides an unfaithful representation of $\hat{\rho}_C$, and is not guaranteed to contain information on the degree of entanglement. If Alice and Bob attempt to evaluate CHSH, the situation is even worse: a basic condition for CHSH is that the state should not change between measurements, and when this condition is broken Alice and Bob can erroneously detect a Bell violation even when Charlie's state is separable. In fact, numerical simulations indicate that separable and entangled sources both violate CHSH with equal probability, $\sim$1 %.

Assuming that $\hat{U}_A$, $\hat{U}_B$ are Haar-random, Alice and Bob know that no particular choice of local measurement basis can give more information than any other. Without loss of generality, we can therefore assume that they *always* measure the $\hat{\sigma}_z$ basis, obtaining expectation values

$$E_j = \langle \hat{\sigma}_z \otimes \hat{\sigma}_z \rangle = \mathrm{Tr}\left((\hat{\sigma}_z \otimes \hat{\sigma}_z)(\hat{\mathcal{U}}_j \hat{\rho}_C \hat{\mathcal{U}}_j^{\dagger}) \right). \tag{4.12}$$

It can easily be shown that the average value of E_j over all $\hat{\mathcal{U}}_j$ is always 0, regardless of $\hat{\rho}_C$. However, if we take the absolute value of E_j before averaging, we will show that the quantity

$$T \equiv \langle |E| \rangle = \sum_{i=0}^{N} \frac{|E_j|}{N}, \tag{4.13}$$

distinguishes entangled states from separable states, and can be used to infer the degree of entanglement of $\hat{\rho}_C$. Figure 4.5c is result of a numerical simulation, showing the distribution of T for a separable state $|00\rangle$ and a maximally entangled state $|\Psi^-\rangle$, averaging over N measurements. All separable pure states give a mean value $T = 1/4$, while all maximally entangled two-qubit states give $T = 1/2$. Partially entangled states give intermediate values. Maximally mixed states give $T = 0$. As the number of averages N is increased, each probability distribution converges towards a Gaussian profile with FWHM proportional to $1/\sqrt{N}$, following the central limit theorem. By taking an increasing number of measurements, Alice and Bob can therefore distinguish a separable state from an entangled state to an arbitrary confidence level.

Proof Given that $\hat{\mathcal{U}} = \hat{U}_A \otimes \hat{U}_B$ where $\hat{U}_{A,B}$ are chosen by the Haar measure on $SU(2)$, we can assume without loss of generality that, after the channel, any separable state is equivalent to the state $|00\rangle$ and any maximally entangled state is equivalent to the singlet $|\Psi^-\rangle$. Unitary rotation of a qubit followed by measurement in the $\hat{\sigma}_z$ basis is equivalent to measurement in the *effective* basis $\hat{M}^{eff} = \hat{U}^\dagger \hat{\sigma}_z \hat{U}$. Rather than integrating $\hat{U}_{A,B}$ over the Haar measure on $SU(2)$, it is simpler to consider the Bloch vectors $\vec{a}_j, \vec{b}_j \in \mathbb{R}^{(3)}$, which depend on $\hat{U}_{A,B}$ and map to Alice and Bob's effective measurement operators

$$\hat{A}_j^{\text{eff}} = \vec{a}_j \cdot \vec{\sigma} = a_j^x \hat{\sigma}_x + a_j^y \hat{\sigma}_y + a_j^z \hat{\sigma}_z \tag{4.14}$$

$$\hat{B}_j^{\text{eff}} = \vec{b}_j \cdot \vec{\sigma} = b_j^x \hat{\sigma}_x + b_j^y \hat{\sigma}_y + b_j^z \hat{\sigma}_z, \tag{4.15}$$

which correspond to points on the 2-sphere $\mathbb{S}^{(2)}$. For the singlet, the expectation value (4.12) is simply given by the dot product, $E_j = -\vec{a}_j \cdot \vec{b}_j$. For $|00\rangle$, the expectation value is $E_j = a_j^z a_j^z$.

In order to compute the average value of T in the asymptotic limit of infinite statistics, we must now integrate $|E|$ over $SU(2) \times SU(2)$. This is equivalent to integrating each Bloch vector over the 2-sphere,

$$\langle |E| \rangle_\infty = \int_{\mathbb{S}^{(2)}} d\vec{a} \int_{\mathbb{S}^{(2)}} d\vec{b}\, |E|. \tag{4.16}$$

First, consider the singlet state. The absolute expectation value $|E| = |\vec{a} \cdot \vec{b}| = |\cos(\phi)|$ depends only on the angle ϕ between $\vec{a}$ and $\vec{b}$. To emphasise, it depends only on the *relationship* between the two channel unitaries. Without loss of generality, we can therefore fix $\vec{a}$ such that $\hat{A}_1 = \hat{\sigma}_z$. Then, we integrate $|E|$ over $\vec{b}$ using a single parameter, ϕ, which rotates $\vec{b}$ about the x-axis of the Bloch sphere. In order to integrate this angle uniformly over $\mathbb{S}^{(2)}$, we must take $\phi = \cos^{-1}(2v - 1)$, where v is uniformly distributed in the interval [0, 1]. Now,

$$\langle |E| \rangle_\infty = \int_0^{2\pi} d\phi |\cos\phi| = \int_0^1 dv |2v - 1| = \frac{1}{2}. \tag{4.17}$$

Now consider the separable state. The absolute expectation value $|E| = |a^z \cdot b^z|$ depends on both the relationship *and* the individual directions of $\vec{a}, \vec{b}$. Writing this expression in terms of the angles ϕ_1, ϕ_2 between $\hat{z}$ and $\vec{a}, \vec{b}$ respectively, we have $|E| = |\cos\phi_1 \cos\phi_2|$. Using the same parametrization to uniformly integrate over $\mathbb{S}^{(2)}$ in (4.16), this becomes

$$\langle |E| \rangle_{\infty} = \int_0^{2\pi} d\phi_1 \int_0^{2\pi} d\phi_2 |\cos\phi_1 \cdot \cos\phi_2| \tag{4.18}$$

$$= \int_0^1 dv_1 \int_0^1 dv_2 |(2v-1)(2v-1)| = \frac{1}{4}. \tag{4.19}$$

So, all maximally entangled two-qubit states have $\bar{T} = 1/2$ and all separable states have $\bar{T} = 1/4$. □

To see how this scheme performs for states other than $|00\rangle$ and $|\Psi^-\rangle$, we consider the state

$$\hat{\rho}_C(v, \mu) = v|\phi(\mu)\rangle\langle\phi(\mu)| + (1-v)\mathbf{1}/4, \tag{4.20}$$

where $|\phi(\mu)\rangle = \sqrt{\mu}|\Psi^-\rangle + \sqrt{1-\mu}|00\rangle$, which can be continuously tuned between a maximally entangled pure state, a separable pure state, and the maximally mixed state. Figure 4.6a shows the results of a numerical calculation of the mean value of T, as a function of the purity and concurrence of $\hat{\rho}_C(v, \mu)$ for various values of v and μ. As we would expect of a sensible entanglement measure, T is maximal for a maximally entangled state ($T = \frac{1}{2}$) and minimal for the maximally mixed state ($T = 0$). As the concurrence or purity of the state is reduced, the strength of correlations is naturally reduced and T falls off monotonically. Note that for separable states, T also gives a measure of purity.

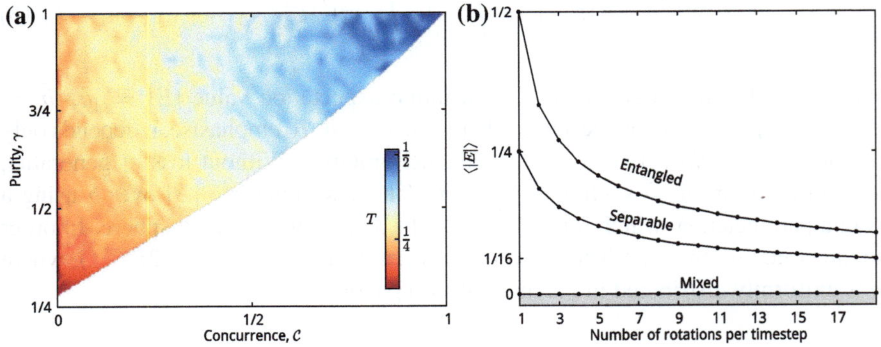

Fig. 4.6 **a** Numerical simulation of average T values for states of varying purity and concurrence. The heatmap shows the mean value of T as a function of the purity $Tr(\hat{\rho}_C^2)$ and concurrence $C(\hat{\rho}_C)$, over states parametrized as $\hat{\rho}_C(v, \mu) = v|\phi(\mu)\rangle\langle\phi(\mu)| + (1-v)\mathbf{1}/4$, with $|\phi(\mu)\rangle = \sqrt{\mu}|\Psi^-\rangle + \sqrt{1-\mu}|00\rangle$. The *white area* of the figure is unphysical: maximally mixed states cannot be maximally entangled. Wavelike features in the figure are an artefact of the numerical interpolation method. **b** Numerical simulation, showing the behaviour of $\langle |E| \rangle$ when the channel fluctuates on a timescale shorter than that required to measure a single expectation value

4.5.1 Experiment

We experimentally tested this scheme using polarization-entangled photon pairs, with both artificial and environmental sources of instability.

4.5.1.1 Experimental Setup

The experimental setup is shown in Fig. 4.5. We used a type-II spontaneous parametric downconversion source, as described in Sect. 1.6.3 of this thesis, to generate entangled photon pairs at 808 nm. A 404 nm Toptica *iBeam* laser at 60 mW was focussed to a waist of $\sim$40 μm on a 2 mm-thick BIBO crystal. We collected down-converted photons at the intersection of the two cones as shown in Fig. 1.8, using two prisms. Each photon was sent through an arrangement of quarter-wave and half-wave plates, allowing arbitrary $SU(2)$ polarization rotations to be applied. A 1 mm-thick uniaxial BIBO crystal was used to compensate for temporal walk-off between horizontal and vertical polarizations. Each arm of the source was then coupled into $\sim$4m of SMF (OZ optics, 808 nm). The measurement setup consisted of two fibre-coupled PBS and four Perkin Elmer APD single-photon detectors, and allows polarization readout in the $\{|H\rangle, |V\rangle\}$ basis. A linear polarizer was optionally inserted into each arm, before the fiber, allowing projective measurements to be performed without the influence of uncontrolled polarization rotations due to the fiber.

4.5.1.2 Source Characterization

The source was optimized to prepare the Bell state

$$|\Psi^+\rangle = \frac{1}{\sqrt{2}}(|01\rangle + |10\rangle) = \frac{1}{\sqrt{2}}(|HV\rangle + |VH\rangle), \tag{4.21}$$

where the phase between $|HV\rangle$ and $|VH\rangle$ terms is determined by the orientation of the compensation crystals. The experimental state was characterized by full quantum state tomography. The waveplates and polarizers shown in Fig. 4.5 were used to implement 36 linearly independent, mutually unbiased measurements, as described for path encoding in Sect. 2.6, and the state was then reconstructed using the same standard maximum-likelihood technique. Real and imaginary parts of the reconstructed density matrix are shown in Fig. 4.7(c, inset). The quantum state fidelity with respect to $|\Psi^-\rangle$ was found to be 0.965 ± 0.002. After losses due to optical elements, fiber coupling, and detector inefficiency, the twofold count-rate registered at the detectors was typically $\sim$1000 counts per second.

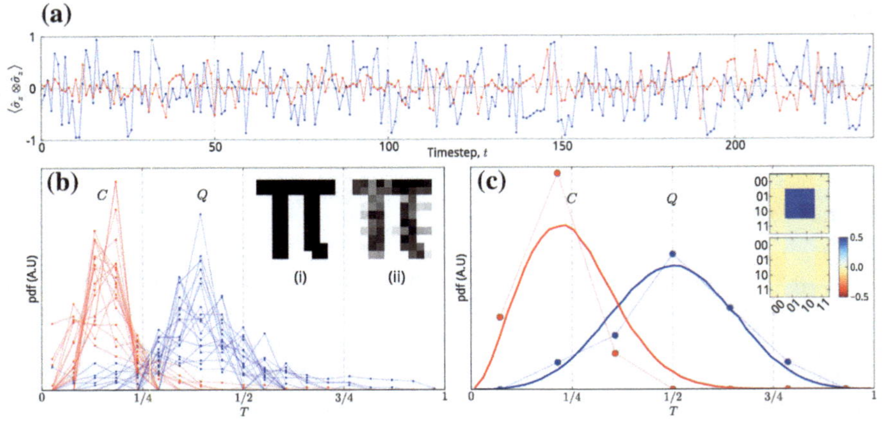

Fig. 4.7 Experimental data. **a** Expectation values $E_j = (C_{00} - C_{01} - C_{10})/\mathcal{C}$, measured as a function of time for a Bell state (*blue lines*) and a separable state (*red lines*), with optical fiber subject to constant bending and twisting. **b** Values of T computed from the data shown in (**a**), with $N \in [1, 20]$. The entangled distributions (*red lines*) are clearly distinguishable from the separable state data (*blue lines*). By encoding bits of information in the choice of entangled/separable state, an image can be sent through the noisy polarization channel. Inset (*i*) source image sent by Charlie, (*ii*) image recovered by Alice and Bob. **c** Twisting optical fiber does not sample uniformly from $SU(2)$. This data was measured using waveplates to experimentally implement ~100 unitaries sampled numerically from $SU(2) \times SU(2)$. *Blue* and *red dots* show the quantum and classical experimental distributions of T respectively, for $N = 4$. *Solid lines* show the theoretical prediction. *Inset* real and imaginary parts of $|\Psi^-\rangle$ as generated by the source, characterized by quantum state tomography

4.5.1.3 Environmental Noise

As already discussed, the polarization of light is not maintained by SMF. The birefringence of the fiber is affected by mechanical stress, and a piece of uniformly stressed fiber has an equivalent effect to a wave-plate, whose characteristic phase-shift depends on the strain, fiber length, core diameter, temperature, and wavelength of light. In fact, *arbitrary* $SU(2)$ polarization rotations can be accomplished using a single piece of SMF, by applying controlled stress to three different regions—these devices are typically marketed as "fiber polarization controllers". A section of fiber exposed to uncontrolled temperature variation and mechanical vibration in the ambient environment of the laboratory will therefore tend to effect a slowly time-varying, arbitrary, random polarization rotation upon the light it carries. The fact that telecom optical fiber networks do not typically use polarization encoding is partly due to cost involved in overcoming this effect.

In our first experiment, we investigated the performance of our technique using random unitary rotations generated in this way. Removing both polarizers, we connected each arm of the source to a fibre-coupled PBS which, together with two detectors, projects onto the $|H\rangle$, $|V\rangle$ states—i.e. measurement in the $\hat{\sigma}_z$ basis. We then recorded coincidence count-rates $c_{HH}, c_{HV}, c_{VH}, c_{VV}$, corresponding to the

$|HH\rangle$, $|HV\rangle$, $|VH\rangle$, $|VV\rangle$ basis states respectively, while manually straining, bending and shaking both optical fibers. We accumulated ~250 expectation values

$$E_j = \frac{c_{HH} - c_{HV} - c_{VH} + c_{VV}}{c_{HH} + c_{HV} + c_{VH} + c_{VV}}. \tag{4.22}$$

Inserting a linear polarizer at 0° into one arm of the source, we then filtered out the $|HV\rangle$ term of (4.21)—rendering Charlie's state separable—and took a second set of data, manipulating the fibers as before. Raw data for both states is shown in Fig. 4.7. Although the average expectation value $\langle E_j \rangle$ is equal to zero for both states, the entangled state is clearly more likely to yield strongly correlated or anticorrelated statistics. Figure 4.7 shows the distribution of the absolute expectation value, when averaging over N measurements, for $N \in [1, 20]$. As predicted, we see a clear distinction between distributions generated by the entangled and separable states, becoming increasingly pronounced with larger values of N. When Charlie prepares $|\Psi^+\rangle$, the average value of T (4.13) was found to be $\langle T \rangle \sim 0.399$. For the separable state $|VH\rangle$, we found $\langle T \rangle \sim 0.163$.

The distribution of random unitaries generated by manual manipulation of SMF is not perfectly uniform. Moreover, it is inevitable that the fiber will move somewhat during each measurement step, in which case the measured expectation value is averaged over a continuum of states. This leads to an overall reduction in the absolute expectation value, and explains why measured values of $\langle T \rangle$ were not closer to 1/2 and 1/4 respectively. Experimentally, it will often be the case that the channel unitary will change by a significant amount during the measurement of a single expectation value. A numerical analysis of the performance of this scheme under such conditions is shown in Fig. 4.6b. Although $\langle T \rangle$ does indeed reduce as the maximum rate of change of the channel is increased, we note a consistent separation between entangled and separable states, suggesting that they might still be distinguished even when the channel fluctuates much faster than a measurement can be made.

In order to illustrate a possible practical application of this scheme, we consider a situation in which Charlie must send a message to Alice and Bob. By switching between entangled and separable state preparation, Charlie can encode the zero and one states of a classical bit, which can then be read out by Alice and Bob—despite noise on the channel. We used this approach to send 100 bits of data, comprising an image of the character π, from source to observer with a statistical fidelity of 85 %. Results are shown in Fig. 4.7(b, inset).

4.5.1.4 Haar-Random Noise

Stressed optical fibre provides a practical example of an unstable environmental channel, but does not typically sample uniformly from $SU(2)$. In order to perform a more controlled test of the theoretical results outlined above, we took measurements using an arrangement of waveplates to implement each qubit channel. Polarizers were inserted into each beam, allowing each qubit to be projected into the $\{|H\rangle, |V\rangle\}$

basis without any influence from the fibre. By setting the fast-axis angles of three consecutive waveplates (QWP, HWP, QWP in that order), any unitary polarization rotation in $SU(2)$ can be realized. Following the approach of Mezzadri [11], we sampled ~40 pseudo-random separable two-qubit unitaries from the Haar measure and solved for the requisite waveplate angles. Setting these angles to Alice and Bob's waveplates, we measured expectation values for both the separable state and the singlet, as before. The distribution of experimentally measured expectation values is shown, for each state, in Fig. 4.7c. We measured mean values of $\langle T \rangle$ of 0.5 ± 0.1 and 0.18 ± 0.07 for the entangled and separable state respectively, compared to ideal theoretical values of 0.5 and 0.25.

4.6 Discussion

The results presented here allow us to formalize a commonly-held, natural and accessible notion of entanglement. Taking two separate systems, a random local operation is applied to each system. Each system is then measured in a local basis. Our main result simply formalizes the fact that, on average, entangled systems yield more strongly correlated measurement outcomes than separable systems. Since our scheme is completely reference-frame independent, we can give this description without speaking of any explicit choice of measurement operators, waveplate angles, or even specific states.

A compelling property of this scheme is the beneficial function of Haar-random, or "white" noise. We see the greatest statistical separation between entangled and separable states, and thus obtain the most information, when instability in the channel is Haar-random. Consider the situation illustrated in Fig. 4.8. Alice and Bob must assess the degree of entanglement of Charlie's state. They are forced to receive qubits

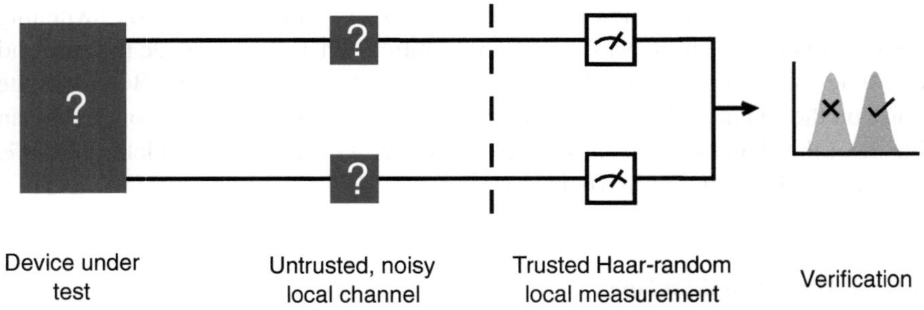

Fig. 4.8 Alice and Bob are tasked with assessing the entangling capability of Charlie's apparatus, but are forced to communicate with him over an untrusted, noisy local channel. By measuring in a controlled Haar-random local basis, they effectively override any fluctuation in the channel, and are able to reliably confirm or deny that Charlie's state is entangled. Even if Charlie has control of the channel, he would need to gain information on Alice and Bob's measurement settings in order to cheat in this scenario

from Charlie over an unstable channel, which is untrusted but guaranteed to be local (i.e. $\hat{\mathcal{U}} \in SU(2) \otimes SU(2)$). In this scenario, there is no guarantee that the channel is Haar-random—it may even be the case that Charlie is deliberately manipulating the channel. However, Alice and Bob can effectively "cancel out" any local operations that may occur on the channel, by *deliberately measuring in a Haar-random basis*. It is then vanishingly unlikely that Alice and Bob will register a large value of T (i.e. $T \sim 0.5$), unless Bob truly has access to an entangled source, or is able to learn Alice and Bob's choice of measurement setting. This ability to override unknown noise on a channel using controlled "white noise", while still obtaining meaningful information on the source, has obvious practical implications for the characterization of quantum states and processes. It would be interesting to consider possible applications outside photonics, where an entangled state must be observed through a noisy local channel.

4.7 Statement of Work

All of the experimental data presented here was measured by myself, and J. Meinecke except for the density matrix in Sect. 4.5. I conceived the original idea in Sect. 4.5. The proof of measurement triads, and Fig. 4.2 are due to my co-authors.

References

1. N. Gisin, G. Ribordy, W. Tittel, H. Zbinden, Quantum cryptography. Rev. Mod. Phys. **74**, 145–195 (2002)
2. J.-Y. Wang, B. Yang, S.-K. Liao, L. Zhang, Q. Shen, X.-F. Hu, J.-C. Wu, S.-J. Yang, H. Jiang, Y.-L. Tang, B. Zhong, H. Liang, W.-Y. Liu, Y.-H. Hu, Y.-M. Huang, B. Qi, J.-G. Ren, G.-S. Pan, J. Yin, J.-J. Jia, Y.-A. Chen, K. Chen, C.-Z. Peng, J.-W. Pan, Direct and full-scale experimental verifications towards ground-satellite quantum key distribution. Nat. Photonics **7**, 387–393 (2013). May
3. R. Ursin, T. Jennewein, J. Kofler, J.M. Perdigues, L. Cacciapuoti, C.J. de Matos, M. Aspelmeyer, A. Valencia, T. Scheidl, A. Acin, C. Barbieri, G. Bianco, C. Brukner, J. Capmany, S. Cova, D. Giggenbach, W. Leeb, R.H. Hadfield, R. Laflamme, N. Lütkenhaus, G. Milburn, M. Peev, T. Ralph, J. Rarity, R. Renner, E. Samain, N. Solomos, W. Tittel, J.P. Torres, M. Toyoshima, A. Ortigosa-Blanch, V. Pruneri, P. Villoresi, I. Walmsley, G. Weihs, H. Weinfurter, M. Zukowski, A. Zeilinger, Space-quest, experiments with quantum entanglement in space. Europhys. News **40**, 26–29 (2009). May
4. P. Zhang, K. Aungskunsiri, E. Martín-López, J. Wabnig, M. Lobino, R.W. Nock, J. Munns, D. Bonneau, P. Jiang, H. W. Li, A. Laing, J.G. Rarity, A.O. Niskanen, M.G. Thompson, J.L. O'Brien, Reference frame independent quantum key distribution server with telecom tether for on-chip client (2013). arXiv:1308.3436
5. C.H. Bennett, D.P. Divincenzo, C.A. Fuchs, T. Mor, E. Rains, P.W. Shor, J.A. Smolin, W.K. Wootters, Quantum nonlocality without entanglement. Phys. Rev. A **59**, 1070–1091 (1999). February
6. Y.C. Liang, N. Harrigan, S. D. Bartlett, T. Rudolph, Nonclassical correlations from randomly chosen local measurements. Phys. Rev. Lett. **104**, 050401 (2010)

7. J.J. Wallman, Y.-C. Liang, S.D. Bartlett, Generating nonclassical correlations without fully aligning measurements. Phys. Rev. A **83**, 022110 (2011). February
8. A. Cabello, Bell's theorem without inequalities and without alignments. Phys. Rev. Lett. **91**, 230403 (2003). December
9. J.J. Wallman, S.D. Bartlett, Observers can always generate nonlocal correlations without aligning measurements by covering all their bases. Phys. Rev. A **85**, 024101 (2012). February
10. D.F.V. James, P.G. Kwiat, W.J. Munro, A.G. White, Measurement of qubits. Phys. Rev. A **64**, 052312 (2001)
11. F. Mezzadri, How to generate random matrices from the classical compact groups. NOTICES of the AMS **54**, 592–604 (2007)
12. J.L. O'Brien, G.J. Pryde, A.G. White, T.C. Ralph, D. Branning, Demonstration of an all-optical quantum controlled-NOT gate. Nature **426**, 264–267 (2003)
13. E.S. Gómez, G. Cañas, J.F. Barra, A. Cabello, Gustavo Lima, Bell tests with random measurements require very high detection efficiencies. Phys. Rev. A **88**, 022102 (2013). Aug
14. M.S. Palsson, J.J. Wallman, A.J. Bennet, G.J. Pryde, Experimentally demonstrating reference-frame-independent violations of bell inequalities. Phys. Rev. A **86**, 032322 (2012). Sep
15. R. Chaves, N. Brunner, Testing nonlocality of a single photon without a shared reference frame. Phys. Rev. A **88**, 012111 (2013)
16. S.M. Tan, D.F. Walls, M.J. Collett, Nonlocality of a single photon. Phys. Rev. Lett. **66**, 252–255 (1991). Jan
17. A. Laing, V. Scarani, J.G. Rarity, J.L. O'Brien, Reference frame independent quantum key distribution. Phys. Rev. A **82**, 012304 (2010)
18. J. Wabnig, D. Bitauld, H.W. Li, A. Laing, J.L. O'Brien, A.O. Niskanen, Demonstration of free-space reference frame independent quantum key distribution. New J. Phys. **15**(7), 073001 (2013). July
19. V. D'Ambrosio, E. Nagali, S.P. Walborn, L. Aolita, S. Slussarenko, L. Marrucci, F. Sciarrino, Complete experimental toolbox for alignment-free quantum communication. Nat. Commun. **3**, 961 (2012)
20. J.A. Slater, C. Branciard, N. Brunner, W. Tittel, Device-dependent and device-independent quantum key distribution without a shared reference frame (2013). arXiv:1311.3343

Chapter 5
Quantum Chemistry on a Photonic Chip

5.1 Introduction

In previous chapters we have seen that quantum mechanics permits strong nonlocal correlations which are classically forbidden. It turns out that this makes it very difficult to engineer a classical digital computer to mimic the behaviour of quantum systems—it seems very likely that the general problem is classically intractable. However, we have good reason to believe that a quantum computer *should* be able to efficiently simulate most quantum systems of interest.

In this chapter we provide a proof-of-principle demonstration of a new algorithm for quantum computers that would give precise calculations of chemical energies and configurations in regimes where classical techniques either fail to give good answers or require exponential computing power. We first examine existing methods for the simulation of quantum systems on classical and quantum computers, with particular focus on quantum chemistry. We then describe our algorithm, and discuss its distinguishing features with respect to existing techniques. Finally, we use this algorithm in a two-photon experiment, simulating the Helium Hydride molecule on the CNOT-MZ chip previously described.

5.2 Simulating Quantum Mechanics

In a large laboratory in Washington DC, a robot arm originally designed to spot-weld car bodies has been installed. Stacked around the walls of the lab are 450,000 micro-test tubes, each containing a different chemical primitive. 24 h a day, seven days a week, this arm, together with a computer vision system, tests prospective drugs for toxicity and efficacy against human-borne diseases [1]. Drug discovery currently has a 99.9 % failure rate, accounting for a significant proportion of the billion dollars it takes to bring a new drug to market. The process of discovery of new high-temperature superconductors, catalysts, and photovoltaics is not far removed from this trial-and-error approach.

P. Shadbolt, *Complexity and Control in Quantum Photonics*,
Springer Theses, DOI 10.1007/978-3-319-21518-1_5

In such fields as mechanical engineering, architecture, microelectronics and aerospace, the design process can be made almost entirely deterministic owing to the power of computer models to predict the success or failure of a given design, without the need for real-world testing. In many cases the computer can itself become the designer, rapidly searching through a large parameter space for optimal geometries or structures. Why is it that many drugs and new materials are not designed in this way?

In *Simulating physics with computers* [2], Feynman describes the intrinsic difficulty of simulating nature, as well as a radical new approach to the problem:

> Nature isn't classical, dammit, and if you want to make a simulation of nature, you'd better make it quantum mechanical, and by golly it's a wonderful problem, because it doesn't look so easy.

Here I will attempt to paraphrase Feynmans argument. Let us define a computer simulation of some physical system as being *efficient* when the number of computer components required (gate operations, memory units and so on) is a polynomial function of the space-time volume of the physical system of interest. If, on the other hand, the necessary computational resources scale exponentially with the problem size, we say that the simulation is inefficient and—if we have any ambition to tackle progressively larger problems—useless. There is generally speaking a one-way correspondence between the space or memory required by an algorithm and its execution time: roughly speaking, if an algorithm really needs an exponential amount of memory, it will not be able to even *address* all that data in polynomial time, let alone solve the problem at hand.

Consider for example the problem of simulating a system of n coins, each of which can be found in the state H or T. The system has 2^n possible states:

$$
\begin{gathered}
H_0 H_1 H_2 H_3 \ldots H_n \\
T_0 H_1 H_2 H_3 \ldots H_n \\
\ldots \\
T_0 T_1 T_2 T_3 \ldots T_n
\end{gathered}
$$

an exponentially large state space. However, the system only ever occupies *one* of these states at a time. Thus the instantaneous state of a system of n coins can always be efficiently represented by n bits of memory, with a simple one-to-one mapping $H \rightarrow 0, T \rightarrow 1$.

Coins flips are often used as a source of randomness. Assuming each flip produces a random output, the expectation value of some function $f(X)$ of n coin flips depends on the probability distribution over all possible outcomes:

$$
\langle f(X)\rangle = \langle f(x_0 x_1 \ldots x_n)\rangle = \sum_{j=1}^{2^n} f(x_j) p(x_j) \tag{5.1}
$$

where x_j is the jth possible outcome of the classical random variable X corresponding to n coin flips. It may appear at first that in order to simulate such a probabilistic system of coins, we must represent the full probability distribution $P(X)$ in the computer's memory, and compute the behaviour of the system by directly evaluating expectation values of the form (5.1). This would again render the problem intractable, since $P(X)$ has exponentially many entries. However, if we allow that the evolution of the computer from state to state can *itself* be random, then we *can* efficiently simulate the physics of coins—simply by exposing bits in memory to a set of probabilistic operations equivalent to those experienced by the coins themselves. In some sense, we generate the probability distribution $P(X)$ without explicitly writing it down. Since the evolution of bits in a deterministic classical computer can be made approximately random with a polynomial overhead in resources, all experiments which depend on random coin flips can be efficiently simulated on a computer.

Now let us consider the problem of simulating a system of n quantum coins, equivalent to spin-$\frac{1}{2}$ particles or qubits. Each coin individually may be in an arbitrary superposition state $|\psi\rangle = \alpha|H\rangle + \beta|T\rangle$. The state of the full system is in general entangled:

$$|\Psi\rangle = a_0|H_0H_1H_2H_3 \dots H_n\rangle \tag{5.2}$$
$$+ a_1|T_0H_1H_2H_3 \dots H_n\rangle \tag{5.3}$$
$$\dots \tag{5.4}$$
$$+ a_{2^n}|T_0T_1T_2T_3 \dots T_n\rangle, \tag{5.5}$$

where a_i are complex probability amplitudes with $\sum_i |a_i^2| = 1$. How should we represent this state on a classical computer? Naïvely, we can write down the real and imaginary parts of each a_i using 2×2^n floating-point variables, an approach which is exponentially costly in time and space. Immediately this representation problem appears hard, but we have previously prevailed in simulating random phenomena, achieving an exponential advantage over the naïve approach through a simple modification of the computer. Can we accomplish a similar trick for quantum coins, and use a classical computer to efficiently represent and evolve the quantum state?[1]

The first piece of evidence to the contrary is the nonlocal behaviour of quantum states, described and experimentally tested in Sects. 1.3.8 and 2.8 and Chap. 4 of this thesis. Since quantum states can exhibit correlations which provably cannot be reproduced by any local classical system, we might expect that it would be difficult to persuade classical bits in a CPU to accurately mimic the evolution and measurement of the quantum state. However, this argument does not say anything about scaling—perhaps such correlations can be emulated, in a completely local way, with a small (polynomial) overhead?

At this point we head into the territory of (quantum) computational complexity theory, where a great deal of beautiful work has been done, but much remains to

---

[1]Note that this question is related to the *Extended Church-Turing Thesis*, discussed in Sect. 6.3.2 of this thesis.

be proved. In 1995, Peter Shor described [3] a polynomial-time quantum algorithm for prime factorization. No polynomial-time classical algorithm for prime factoring exists, and the problem is generally believed to be exponentially hard for classical computers, although there is no proof. If factoring is indeed outside of **P**, then a universal full-scale quantum computer running Shor's algorithm would constitute a "physical system of interest", albeit contrived, which cannot be efficiently simulated by any classical machine. Further evidence has recently been provided by Aaronson and Arkhipov, in their resent proposal for the BOSONSAMPLING linear optical quantum computer, discussed in detail in Sect. 6.3.2. The authors provide very strong evidence that efficient simulation of the quantum behaviour of single photons in certain classes of linear optical network is classically intractable. This result arguably has stronger implications for the complexity of quantum simulation, as the implications of a polynomial-time classical algorithm for BOSONSAMPLING would be much more dramatic than the discovery of a fast classical factoring algorithm.

So we end up with a reasonable hunch that the simulation of small things—molecules, drugs, materials—is sometimes classically intractable, and we can see a number of bright lights in the darkness which support this understanding. This is not to say that *all* quantum systems are intrinsically difficult to simulate classically, for instance, an n-body system whose state remains separable throughout its evolution is simulated using the same method as for probabilistic classical systems. Only a subset of natural phenomena exhibit sufficiently strong quantum correlations as to be classically intractable. Certain regimes of organic [4] and inorganic chemistry [5], superconducting materials [6, 7] and quantum magnetism [8], and microbiology, for instance photosynthesis [9], all fall into this regime. Here we will focus our attention on problems in the field of *quantum chemistry*.

5.3 Quantum Chemistry

> The underlying physical laws necessary for the mathematical theory of a large part of physics and the whole of chemistry are thus completely known, and the difficulty is only that the exact application of these laws leads to equations much too complicated to be soluble. [10]
>
> Paul Dirac, 1929

Quantum chemistry is the experimental and theoretical study of the quantum mechanical behaviour of chemicals. The fundamental goal is the ability to compute and comprehend the properties and dynamics of large molecules, without the need to directly synthesise and test them in the lab. Owing to the complexity of these calculations, a considerable fraction of this research is dedicated to numerical studies. The roots of the field lie in the early observations of quantum electronic behaviour due to Faraday, Kirchhoff, Boltzmann and Planck. Later developments were made by Linus Pauling, in his famous work on the quantum mechanical nature of the chemical bond [11], as well as Llewellyn Thomas and Enrico Fermi, to name but a few.

5.3.1 *Definition of the Problem*

Let's assume that we know the chemical composition of a molecule of interest, having some information on its geometry, the relative positions, masses and charges of the nuclei, etc. For most chemical systems of interest, the full molecular wavefunction Ψ can be factorized into electronic and nuclear components via the Born-Oppenheimer approximation [12]

$$\Psi = \psi_e \times \psi_n, \tag{5.6}$$

after which we assume that the nuclei are stationary and effectively classical, since they are so much more massive than the electron. The problem then is to solve the time-independent Schrödinger equation for a system of N nuclei and n electrons

$$i\hbar\frac{\partial}{\partial t}\psi_e = \hat{H}_e\psi_e, \tag{5.7}$$

where $\hat{H}_e$ is the Hamiltonian for the electronic structure problem, which can be written [13] in second-quantized form as

$$\hat{H}_e = \sum_{ij} h_{ij}\hat{a}_i^\dagger\hat{a}_j + \sum_{ijab} h_{ijab}\hat{a}_i^\dagger\hat{a}_j^\dagger\hat{a}_a\hat{a}_b. \tag{5.8}$$

Here $\hat{a}_j^\dagger$ and $\hat{a}_j$ are the fermionic ladder operators, which create and destroy electrons in a *molecular spin orbital* ("energy level") j. The first term in (5.8) is due to the electronic kinetic energy, the second is a result of electron-electron (Coulomb) interaction.

Analytic solutions to the electronic structure problem exist for small molecules such as the Hydrogen atom, but in general we must take a numerical approach. The basic quantity of interest for chemists is usually an energy $E = \langle\lambda_0|\hat{H}_e|\lambda_0\rangle$ or energy difference ΔE, where $|\lambda_0\rangle$ is an eigenstate of $\hat{H}$. Frequently we are interested in the dependence of this energy on some molecular or external degree of freedom:

- How much effort must we exert in order to pull this atom away from the rest of the molecule? What is the complete form of the interaction potential energy surface of the molecule as a function of its own configuration?
- How high is the energy barrier that we must overcome in order to persuade two molecules of interest to react?
- How stable is this compound? How much energy would it take to pull it apart?

An example of a very simple approximate solution to such questions is the Lennard-Jones potential,

$$V_{LJ} = \varepsilon\left[\left(\frac{r_m}{r}\right)^{12} - 2\left(\frac{r_m}{r}\right)^6\right] \tag{5.9}$$

which approximates the dependence of the interaction potential on the distance r between two atoms, where ϵ is the depth of the potential well at $r = r_m$, the equilibrium bond length of the molecule. Lennard-Jones gives a simple and computationally frugal estimate of the interaction energy, but its approximation breaks down for a broad variety of chemical systems. For larger, more complex molecules, quantum chemists depend on more sophisticated models, or *ansätze*.

5.3.2 Ansätze

The first task in solving problems of the form of (5.7) is to choose a representation, parametrization or *ansatz* for the electronic wavefunction Ψ_e. The *molecular orbital approximation* gives a simple ansatz for the molecular electronic structure, in which the full electronic wavefunction Ψ is written as a separable product of single-electron molecular wavefunctions ψ_i:

$$\Psi(\vec{r}_1, \vec{r}_2, \ldots \vec{r}_n) = \prod_{i=1}^{N} \psi_i(\vec{r}_i). \tag{5.10}$$

known as a Hartree product. Any single-electron molecular wavefunction can be expressed as a linear combination over a basis set of n_{basis} atomic orbitals (single-electron, single-atom wavefunctions) ϕ_j,

$$\psi_i(\vec{r}) = \sum_{j=1}^{n_{basis}} c_{ij}\phi_j(\vec{r}). \tag{5.11}$$

In general, the Hartree product (5.10) violates Pauli exclusion, since it is not antisymmetric: the expressions $\psi(\vec{r}_1, \vec{r}_2) = \psi(\vec{r}_1) \times \psi(\vec{r}_1)$ and $\psi(\vec{r}_2, \vec{r}_1) = \psi(\vec{r}_2) \times \psi(\vec{r}_1)$ are not the same, and

$$\psi(\vec{r}_1, \vec{r}_2) \neq -\psi(\vec{r}_2, \vec{r}_1), \tag{5.12}$$

i.e. the electronic wavefunction does not change sign upon exchange of two electrons. The solution is to antisymmetrize the wavefunction, writing it as a linear combination of Hartree products

$$\psi(\vec{r}_1\vec{r}_2) = \frac{1}{\sqrt{2}}\left(\psi_1(\vec{r}_1)\psi_2(\vec{r}_2) - \psi_2(\vec{r}_2)\psi_1(\vec{r}_1)\right). \tag{5.13}$$

Using a method due to Slater [14], we can generalize this ansatz to the n-electron case, including the electron spin, by writing the full electron wavefunction as an antisymmetrised ($\mathcal{A}$) product of spin orbitals $\chi_i(\vec{r}_i, \omega) \in \left[\psi_i(\vec{r}_i)\alpha(\uparrow), \psi_i(\vec{r}_i)\beta(\downarrow)\right]$,

$$\Psi(\vec{r}, \omega) \equiv \Psi(x) = \mathcal{A}\{\prod_{i=1}^{n} \chi_i(x_i)\}, \tag{5.14}$$

which can be neatly written as a *Slater determinant*

$$\Psi(x_1, x_2 \ldots x_n) = |\chi_1 \chi_2 \ldots \chi_n\rangle = \frac{1}{\sqrt{n!}} \begin{vmatrix} \chi_1(x_1) & \chi_2(x_1) & \ldots & \chi_n(x_1) \\ \chi_1(x_2) & \chi_2(x_2) & \ldots & \chi_n(x_2) \\ \vdots & \vdots & \ddots & \vdots \\ \chi_1(x_n) & \chi_2(x_n) & \ldots & \chi_n(x_n) \end{vmatrix}. \tag{5.15}$$

The Slater determinant provides an elegant ansatz for separable molecular spin orbitals, which is physical by construction. Note that this is the fermionic equivalent of the method described in Sect. 1.5.3 to compute bosonic states and statistics using the permanent $\text{per}(M)$. Owing to the fact that states described in this way do not include any entanglement, *every state in the ansatz can be parametrized with a polynomial number of parameters*, the single-electron atomic orbital coefficients c_{ij} in (5.11). We will herein label the real parameters used to address such a subspace of states as $\vec{\phi} \equiv c_{ij}$.

5.3.2.1 Hartree-Fock

Having chosen an ansatz for the state, the task is then to find the parameter values $\vec{\phi}$ which best satisfy the Schrödinger equation. The *variational principle* states that any trial wavefunction (a "guess" at $\vec{\phi}$) will not have an energy less than the ground state energy E_0 of the Hamiltonian. Therefore we can find a good, approximate solution to the Schrödinger equation—the ground state itself—simply by varying these parameters so as to minimize the energy, in what is known as the *variational method.* This technique lends itself to a numerical approach, in which an iterative nonlinear optimization algorithm is used to minimize the energy of a trial wavefunction,

$$E_0 = \min_{\vec{\phi}} \langle \Psi(\vec{\phi}) | \hat{H} | \Psi(\vec{\phi}) \rangle. \tag{5.16}$$

From $\vec{\phi}$, we can then reconstruct full (approximate) information of the electronic configuration, as well as the ground state energy E_0.

The *Hartree-Fock-Roothan (HF) method* is an iterative, polynomial-time algorithm which computes an approximate solution to (5.16), yielding the HF ground state $|\Phi_0\rangle$. Key to the efficiency of this technique are two related assumptions: (i) that Ψ is separable, allowing it to be expressed as a single Slater determinant, and (ii) that the Coulomb interaction term in $\hat{H}_e$ is well-described by a *mean-field approximation* in which all two-electron contributions are approximated "as well as possible" by single-electron terms in the same Slater determinant.

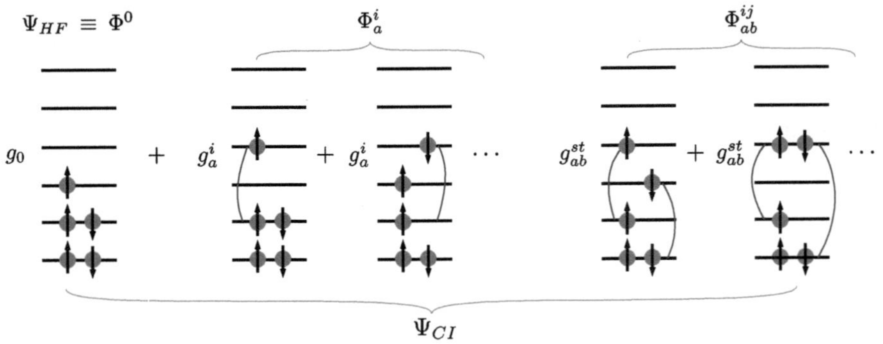

Fig. 5.1 Schematic of the configuration interaction ansatz—a linear combination of possible molecular spin orbital configurations. When the series is not truncated, we obtain the *full configuration interaction* ansatz, which is exact up to the choice of atomic orbital basis set. However, given the number of orbitals and electrons in a typical molecule of interest, and the number of permutations thereof, this encoding is classically intractable for systems of more than $\gtrsim 3$ atoms

HF provides a polynomial ansatz which has been very successful in describing a broad range of chemical systems, but does not account for electron correlations or nonseparability. As such this method fails for many physical systems of interest, including those described in the introduction to this chapter. In an attempt to remedy this situation, correlated electronic behaviour has been re-introduced to the ansatz by a number of "post-Hartree-Fock" methods.

5.3.2.2 Post Hartree-Fock

In the Hartree-Fock method, the electronic configuration wavefunction is approximately parametrized in terms of a single Slater determinant. A numerically exact, unscalable, completely general ansatz is given by the fully configuration interaction(FCI) method, illustrated in Fig. 5.1, in which the entire space of physical electronic wavefunctions is fully and exactly parametrized using a linear combination of exponentially many Slater determinants, accounting for all possible (entangled, correlated) electronic configurations

$$\Phi_0 = |\chi_1 \chi_2 \chi_3 \cdots \chi_n\rangle \tag{5.17}$$

$$\Psi_{CI} = g_0 \Phi^0 + \sum_{a,i} g_a^i \Phi_a^i + \sum_{a,b,i,j} g_{ab}^{st} \Phi_{ab}^{ij} \cdots \tag{5.18}$$

where the spin-orbital subscripts $(a, b \ldots)$ and superscripts $(i, j \ldots)$ mark differences with respect to the Hartree-Fock ground state. FCI calculations give numerically exact, optimal solutions, but the number of Slater determinants, and thus the number of parameters required to describe the state, scales factorially with the number of electrons. As such, FCI calculations are currently limited to diatomic or triatomic molecules.

Strongly related to CI methods is the *coupled-cluster* (CC) ansatz [15]. Configuration-interaction methods can be made tractable by truncation of the series (5.18). CC methods provide an improved approach to this truncation, grouping electronic excitations together in the exponential ansatz,

$$|\Psi\rangle = e^{\hat{T}}|\Phi_0\rangle. \tag{5.19}$$

Here, $|\Phi_0\rangle$ is the Hartree-Fock ground state, which can be efficiently computed as we have already seen, and $\hat{T}$ is the so-called *cluster operator*. The basic technique is to group k-fold electronic excitations, choosing a cut-off at $k = k_{max}$:

$$T = T_1 + T_2 + T_3 \ldots T_{n_{max}} \tag{5.20}$$

where T_1 is the single-excitation term, a linear combination of all possible excitations which raise or lower a single electron from spin orbital a to i,

$$T_1 = \sum_i \sum_a g_a^i \hat{a}^a \hat{a}_i^\dagger. \tag{5.21}$$

The pair excitation term is more complex, simultaneously raising two electrons from spin orbitals (a, b) to (i, j)

$$T_2 = \frac{1}{4} \sum_{ij} \sum_{ab} g_{ab}^{ij} \hat{a}^a \hat{a}^b \hat{a}_i^\dagger \hat{a}_j^\dagger \tag{5.22}$$

and so on. In practice, this series is usually truncated at the level of two-particle or three-particle excitations. By this approximation, the number of parameters used to describe the state remains polynomial in the system size. Even so, the CC ansatz is currently classically intractable for $k_{max} \gtrsim 3$.

We will skip discussion of *density functional theory* (DFT), an alternative mean-field theory for quantum chemistry (see [16]). Suffice to say that despite the success of DFT, as with the HF, CC and truncated CI methods, the approximation that it uses to achieve scalability leads to incorrect results for a large class of chemical systems.

5.4 Quantum Simulators

We have arrived a situation in which all known exact methods for the simulation of quantum chemistry are intractable for molecules with more than ~3 atoms. Moreover, the approximate methods that do scale are only precise for certain classes of molecule. Hartree-Fock, coupled-cluster, DFT and truncated CI models all break down at some point. There are examples of surprisingly simple molecules for which all known approximate methods fail, including the lowly nitrogen N_2 molecule, whose triple bond gives rise to strongly correlated electronic behaviour at high bond separations, ozone, and many others. How should we go about simulating these systems and their larger, more interesting cousins?

If we are serious about efficient simulation of quantum mechanical phenomena in the lab, then the computer or machine that we use must also be quantum mechanical—this was Feynman's insight. Throughout his work, Feynman acknowledged the possibility that the device might not necessarily constitute a universal full-scale quantum computer. We can imagine a broad variety of special purpose devices, which perhaps do not even depend on digital quantum logic or gate operations, but nonetheless emulate or mimic the physics of a classically intractable system of interest in a scalable way. The potential for dramatic relaxation of hardware requirements (in terms of coherence time, gate fidelity etc.) in this regime, while maintaining a quantum advantage, has led many to predict that non-universal quantum simulation may constitute the first practical application of large-scale artificial quantum entanglement.

In this chapter we will only discuss schemes for quantum chemistry which *do* make use of a full-scale universal digital quantum computer, and we do not address special-purpose, non-universal devices. See the discussion of BOSONSAMPLING in Sect. 6.3.2 of this thesis for an experimental and theoretical examination of special-purpose quantum simulators, as well as recent experimental progress in [8, 17].

5.4.1 Quantum Simulation on a Digital Quantum Computer

We will now give a picture of the standard approach to quantum simulation on a universal digital quantum computer. An enormous diversity of methods exist, and this description will necessarily be approximate and incomplete. We will later compare and contrast this standard method with the technique used in our experiment, which is quite distinct.

In any computer simulation, we must choose a mapping between the degrees of freedom of the physical system of interest and the computational hardware. We have already seen the approach taken in classical quantum chemistry, in which an ansatz for the electronic wavefunction is expressed in terms of atomic spin orbitals, the coefficients of which are stored as floating-point numbers in a digital register. In a quantum computer, quantum information is written into registers of qubits—distinguishable spin-$1/2$ systems. Onto this register we wish to encode the state of a system of n electrons—indistinguishable, antisymmetric fermions, with half-integer spin. In his original discussion of universal quantum simulators, Feynman expressed concern over the discrepancy between the fundamental physical properties of these two systems [2]. How should we reconcile the two?

5.4.1.1 The Jordan-Wigner Transform

Suppose that we have register of N qubits, onto which we would like to map the state of n electrons. We can dream up many possible encodings, but most of them will allow us to create or destroy simulated electrons in unphysical ways. For example, we should not be able create two electrons occupying the same spin orbital, and annihilation on the vacuum should produce no effect. The essential rules for the fermionic creation and annihilation operators acting on a mode j are completely captured by the (fermionic) *canonical anticommutation relations* (fCCRs):

$$\{\hat{a}_j, \hat{a}_k^\dagger\} = \delta_{jk} I \tag{5.23}$$
$$\{\hat{a}_j, \hat{a}_k\} = 0 \tag{5.24}$$

where $\{A, B\} = AB + BA$ is the anticommutator [18]. These equations, which are the fermionic counterpart to the bosonic CCRs (1.101) immediately imply that $\{\hat{a}_j^\dagger, \hat{a}_k^\dagger\} = 0$, and $(\hat{a}_j^\dagger)^2 = (\hat{a}_j)^2 = 0$, the $\hat{a}^\dagger \hat{a}$ are positive Hermitian with eigenvalues 0 and 1 and are mutually commuting $\hat{a}_j^\dagger \hat{a}_j \hat{a}_k^\dagger \hat{a}_k = \hat{a}_k^\dagger \hat{a}_k \hat{a}_j^\dagger \hat{a}_j$, and annihilation on the vacuum behaves as desired ($\hat{a}|0\rangle = 0$).

The *Jordan-Wigner transform* [19, 20] provides exactly such a mapping from qubits to fermions in the form of a definition for $\hat{a}^\dagger, \hat{a}$ in terms of spin operators acting on qubits which always satisfies the fCCRs. The Jordan-Wigner transform allows *any physical system* to be represented on a quantum computer, and thus forms the basic ingredient for the encodings used in most digital quantum simulation algorithms [21, 22].

In terms of the Pauli spin operators $\hat{\sigma}_i$, the fermionic creation and annihilation operators acting on mode j are defined by Jordan-Wigner as

$$\hat{a}_j \equiv I^{\otimes j-1} \otimes \hat{\sigma}_+ \otimes \hat{\sigma}_z^{\otimes N-j} \tag{5.25}$$
$$\hat{a}_j^\dagger \equiv I^{\otimes j-1} \otimes \hat{\sigma}_- \otimes \hat{\sigma}_z^{\otimes N-j} \tag{5.26}$$

where $\sigma_+ = |0\rangle\langle 1|$ and $\sigma_- = |1\rangle\langle 0|$. The tall stack of z-rotations ($\hat{\sigma}_z^{\otimes n-j}$, sometimes referred to as Jordan-Wigner *ladder*) has has the effect of keeping track of the sign of the fermionic wavefunction and thus enforcing antisymmetry:

$$\hat{a}_j |\alpha_1, \alpha_2, \ldots \alpha_l\rangle = -(-1)^{s_j^\alpha} |\alpha_1, \alpha_2, \ldots \alpha_l, \text{ with } \alpha_j \rightarrow 0\rangle, \tag{5.27}$$

where $|\alpha_1, \alpha_2 \ldots \alpha_n\rangle$ is the occupation number representation of the fermionic state and $s_j^\alpha \equiv \sum_{k=1}^{j-1} \alpha_k$. It is interesting to note that when we make a *local* change to the electronic system—creating, annihilating or moving an electron—the corresponding qubit operator, i.e. the necessary gate operation, is highly *nonlocal*.

5.4.1.2 Quantum Phase Estimation

Having mapped the physics of the chemical system into a digital register of qubits, the task is then to design a quantum circuit—a sequence of gate operations—which computes the eigenenergies of the electronic structure Hamiltonian of interest. Here we provide an approximate picture of the traditional framework, which is based on the quantum phase-estimation algorithm (PEA) [23, 24].

The PEA takes as input an eigenstate $|\lambda_0\rangle$ of a unitary operator $\hat{U}$, and computes a t-bit approximation to the unknown phase φ_{λ_0} of the eigenvalue $\lambda_0 = e^{2\pi i \varphi_{\lambda_0}}$. PEA is an oracle-based algorithm, and starts from the assumption that the controlled-$\hat{U}^{2^j}$ operation can be implemented by a black-box, for arbitrary j, at a *constant* cost. The controlled-unitary gates act as

$$C(\hat{U}^{2^k})\,|+\rangle \otimes |\lambda_0\rangle = \frac{1}{\sqrt{2}}\left(|0\rangle + e^{2\pi i \cdot 2^k \varphi}\,|1\rangle\right) \otimes |\lambda_0\rangle. \tag{5.28}$$

The system register is first initialized in the eigenstate $|\lambda_0\rangle$, which is provided as input to the algorithm. A secondary control register of t qubits is prepared in the separable equal superposition state $|+\rangle^{\otimes t}$. We then apply the circuit of controlled-$\hat{U}^{2^t}$ operations shown in Fig. 5.2. The system register stays in the state $|\lambda_0\rangle$ throughout the computation, with the full system evolving as

$$\begin{aligned} |+\rangle^{\otimes t} \otimes |\lambda_0\rangle \xrightarrow{\text{PEA}_1} \frac{1}{\sqrt{2^0}} &\left(|0\rangle + e^{2\pi i \cdot 2^{t-1}\varphi}|1\rangle\right) \\ &\otimes \left(|0\rangle + e^{2\pi i \cdot 2^{t-2}\varphi}|1\rangle\right) \\ &\cdots \\ &\otimes \left(|0\rangle + e^{2\pi i \cdot 2^{0}\varphi}|1\rangle\right) \otimes |\lambda_0\rangle. \end{aligned} \tag{5.29}$$

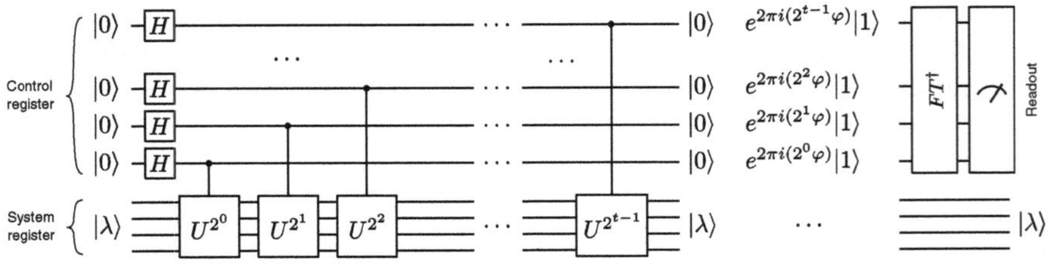

Fig. 5.2 The PEA computes an t-bit approximation to the phase φ_{λ_0} of the eigenvalue $\lambda_0 = e^{2\pi i \varphi_{\lambda_0}}$ of a unitary operator, U, assuming that the eigenstate $|\lambda_0\rangle$ is given. If arbitrary exponentiation U^{2^j} up to U^{2^t} is provided as a *black-box* oracle, then the PEA can achieve an exponential speedup over classical methods. The eigenstate is prepared in the *system register*, and the *control register* of t qubits is prepared in the superposition state $|+\rangle^{\otimes t}$. The system evolves under repeated application of the oracle unitary, quantum-controlled by qubits in the control register. Finally, *readout* of φ_{λ_0} is performed by means of the inverse quantum Fourier transform followed by measurement in the computational basis

The final state of the control register can then be written independently of the system,

$$\frac{1}{\sqrt{2^t}}\left[|00\ldots0\rangle + e^{2\pi i\varphi\cdot 1}|00\ldots1\rangle + \ldots e^{2\pi i\varphi\cdot 2^{t-1}}|11\ldots1\rangle\right] \tag{5.30}$$

$$= \frac{1}{\sqrt{2^t}}\sum_{k=0}^{2^t-1} e^{2\pi i\cdot\varphi k}|k\rangle \tag{5.31}$$

where $|k\rangle$ is the state corresponding to the binary representation of k. In the event that the phase can be exactly written as a binary fraction of t bits $\varphi = 0.\varphi_1\varphi_2\ldots\varphi_t \equiv \frac{\varphi_1}{2} + \frac{\varphi_2}{4} + \frac{\varphi_3}{8}\ldots\frac{\varphi_t}{2^t}$, the output state of the first stage of PEA (5.29) can be rewritten as

$$\frac{1}{\sqrt{2^t}}\left(|0\rangle + e^{2\pi i\,0.\varphi_t}|1\rangle\right)\otimes\left(|0\rangle + e^{2\pi i\,0.\varphi_{t-1}}|1\rangle\right)\ldots\otimes\left(|0\rangle + e^{2\pi i\,0.\varphi_1\varphi_2\ldots\varphi_t}|1\rangle\right) \tag{5.32}$$

It is then straightforward to show that the quantum Fourier transform of (5.32) is a logical basis state corresponding to the digits of φ, $|\varphi_1\varphi_2\ldots\varphi_t\rangle$. The final stage of the PEA implements this quantum Fourier transform on the control register, followed by measurement in the logical basis. From these measurement outcomes the experimentalist reads out the exact digits of φ, thereby obtaining the eigenvalue λ_0 of $\hat{U}$ in a single shot. Even when φ cannot be exactly expressed as a t-bit binary fraction, the PEA returns the phase to a good approximation, with a success probability $1-\epsilon$. The choice of t determines the output precision as well as the probability of success of the PEA.

5.4.1.3 Quantum Chemistry Using the PEA

We will now describe a polynomial-time algorithm which makes use of the PEA to compute exact ground-state energies under the full configuration-interaction ansatz, following [25, 26]. Starting from the FCI electronic structure Hamiltonian $\hat{H}$ (5.8) for our molecule of interest, we generate the unitary time evolution operator $\hat{U} = e^{i\hat{H}\tau}$, where the energy $E = 2\pi\varphi/\tau$ of an eigenstate $|\lambda_0\rangle$ is mapped to the phase of its eigenvalue $\lambda_0 = e^{2\pi i\varphi}$

$$\hat{U}|\psi\rangle = e^{i\hat{H}\tau}|\psi\rangle = e^{2\pi i\varphi}|\psi\rangle. \tag{5.33}$$

The task is then to estimate φ, and thus E, by means of the PEA. We must first be able to implement the controlled-$\hat{U}(\tau)$ operations at the heart of the PEA as gate sequences in a digital quantum computer. The Trotter decomposition provides a general prescription for approximate time evolution of arbitrary unitary operators in the gate model. The technique bears a strong resemblance to stop-frame animation

[27]. For a Hamiltonian $\hat{H} = \sum_k h_k$, the full time-evolution $\exp(i\hat{H}\tau)$ is divided into M short, time-independent unitary slices of length $\Delta_\tau = \tau/M$,

$$\hat{U} = e^{i\tau\sum_k h_k} = \prod_k \left[e^{i\Delta_\tau h_k}\right] + O(\Delta_\tau), \tag{5.34}$$

a process known as *Trotterization*. The number of gate operations is at least linear in t, and the procedure introduces a discretization error polynomial in t. Larger values of M and higher-order decompositions both give rise to a smoother "animation" and less error, at the cost of further gate operations. This error must be within chemical accuracy (roughly one part in a million) for the computation to be useful. Even for small molecules, Trotterization to chemical accuracy demands very large numbers of gates. A conservative implementation of FCI-PEA for the water molecule using ~30 qubits in the system register requires $O(10^4)$ gate operations per Trotter step, and $M = O(10^6)$ steps in the full time evolution, leading to a total of $O(10^{10})$ serial gate operations [26]—a formidable challenge.

The PEA provides an efficient method to compute the eigenvalue of a *given* eigenstate of $\hat{U}$. However, in the context of quantum chemistry we do not initially know the eigenstate—in fact, $|\lambda_0\rangle$ should be regarded as encoding the *answer* to our problem. Efficient classical methods provide an approximate ground state, but the error in this approximation is the entire motivation to seek a quantum algorithm in the first place! What happens if we use the approximate Hartree-Fock ground state $|\psi_{HF}\rangle$ instead of the exact eigenstate? Using $|\psi_{HF}\rangle = \sum_i \lambda_i|\lambda_i\rangle \approx |\lambda_0\rangle$ as input to the system register, we find [24] that the PEA outputs the exact phase φ of $|\lambda_0\rangle$ with probability proportional to $|\langle\lambda_0|\psi_{HF}\rangle|^2$. Thus in some cases, an approximate eigenstate can be used to find the *exact* energy of the ground state, at the cost of a lower probability of success.

This method has been used for ground state estimation in numerical simulations of H_2O [26, 28] and LiH [28], as well as a recent experimental demonstration for H_2 using photonic qubits [29]. Unfortunately, for many chemical systems of interest, the Hartree-Fock approximation performs so badly that the probability of success vanishes. As a result, the state preparation problem in quantum simulation of quantum chemistry can become very involved. Methods to overcome this issue largely depend on adiabatic eigenstate preparation algorithms [28, 30] in which the Hamiltonian is slowly transformed from an "easy" Hartree-Fock Hamiltonian $\hat{H}^{HF}$ to the exact, full-configuration interaction Hamiltonian $\hat{H}^{FCI}$. These methods again depend on Trotterization for the implementation of time-evolution under a time-dependent Hamiltonian, incurring a similar or greater cost in the number of required gate operations.

We thus arrive at a *ballistic* picture of quantum algorithms for quantum chemistry resembling that shown in Fig. 5.3, in which the process is broadly subdivided into (i) preparation of qubits in a simple fiducial state $|0\rangle^{\otimes N}$ (ii) adiabatic or iterative PEA state preparation and (iii) PEA readout of the energy. A key property of this approach is that while the number of qubits N can be relatively small—PEA is

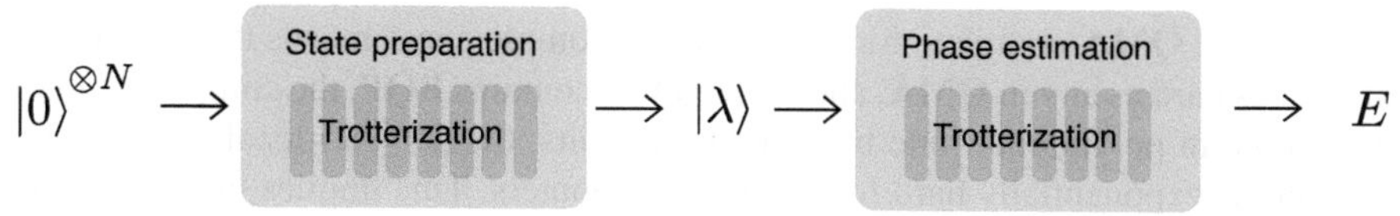

Fig. 5.3 Ballistic quantum chemistry on a quantum computer. A fiducial state $|0^{\otimes N}\rangle$ is adiabatically time-evolved to an eigenstate $|\lambda_0\rangle$ of the Hamiltonian of interest. The energy is then read out by means of the quantum phase estimation algorithm. A significant property of this approach is that although the necessary number of qubits can be relatively low, the number of fundamental gate operations which must be consecutively and coherently performed is typically very large due to the heavy dependence on Trotterization for time-evolution of the state

amenable to a recursive modification which allows chemically relevant calculations to be performed using $\sim$10 control qubits and $\sim$30 system qubits—the number of basic gate operations required is typically enormous.

5.4.2 Limitations of Quantum Simulators

Arguably the most important task for in any scalable algorithm for quantum chemistry is the choice of ansatz. The most general ansatz, which captures the full space of possible states of the system and maps to the full *Hilbert space* $\mathscr{H}$ upon which the quantum state is defined has dimension $O(2^n)$ and can only be parametrized by an exponential number of real parameters. Efficient classical algorithms must therefore *throw away* an exponentially large subspace of $\mathscr{H}$. The most successful ansätze do this in a targeted way, discarding highly entangled or extremely strongly correlated states—which do not often appear in nature—while preserving the most chemically relevant regions of $\mathscr{H}$.

The Hilbert space dimension of n qubits and that of n electrons occupying a system of spin orbitals are both exponential in n and are of the same order. We have seen from the Jordan-Wigner transform that the physics of these two systems can be made isomorphic, and from this it might be natural to infer that a quantum computer should be able to implement a *complete* ansatz, addressing the entirety of $\mathscr{H}$. The counter-argument to this reasoning is simple: we need to be able to *drive* the quantum computer. That is, any machine which allows us to prepare or represent states throughout the entirety of $\mathscr{H}$ must by definition have a number of classical control parameters—knobs—exponential in n, and is therefore not scalable. Quantum computers must have a polynomial number of knobs on top, and as such can only access a polynomially small subset of *efficiently preparable* Hilbert space. *Arbitrary n-qubit state preparation does not scale.*

Ground-state quantum chemistry problems are a subset of the k-local Hamiltonian problem, i.e. the problem of finding the ground state of a Hamiltonian on n qubits, $\hat{H} = \sum i = 1^r \hat{h}_i$ where $r = \text{poly}(n)$ and each $\hat{h}_i$ acts on at most k qubits. The general k-local Hamiltonian problem has been proven [31] by Kempe, Kitaev and

Regev to be **QMA**-complete for $k \geq 2$. **QMA**-completeness means that problem is at least as hard as any in **QMA**, and since **QMA** contains **BQP**, the complexity class accessible in polynomial time by quantum circuits, the $(k \geq 2)$-local Hamiltonian problem is exponentially hard for quantum computers. This implies that there exist polynomial-size ground state problems in quantum chemistry that are intractable *even on a quantum computer*.

If all of the above is true, why should we bother to build a quantum computer for quantum chemistry? Despite the apparent difficulty of building a digital quantum simulator, a small fraction of which we have outlined outlined above, we nonetheless expect that such devices should provide an exponential speedup over classical machines for large classes of interesting physical and chemical systems, enabling FCI quantum chemistry in polynomial time. That said, it would be nice if we could do it without the need for *quite* so many gate operations. The next section presents our work in this direction.

5.5 Quantum Simulation Without Quantum Evolution

We will now describe an alternative approach by which quantum chemistry calculations can be performed on a hybrid quantum-classical processor without time evolution or quantum phase estimation. This approach introduces a number of new unknowns, but significantly reduces the number of required gate operations by means of a variational approach with a strong resemblance to certain classical methods in quantum chemistry.

5.5.1 Scheme

Any Hamiltonian can be written as a sum of tensor products of Pauli matrices

$$\hat{H} = \sum_{i\alpha} h^i_a \hat{\sigma}^i_a + \sum_{ijab} h^{ij}_{ab} \hat{\sigma}^i_a \otimes \hat{\sigma}^j_b + \cdots \tag{5.35}$$

for real h, where $(a, b\ldots)$ index the three Pauli operators $\{\hat{\sigma}_x, \hat{\sigma}_y, \hat{\sigma}_z\}$ and $(i, j\ldots)$ index the subspace of qubits upon which they act. In general this expansion has exponentially many terms, but for all physical Hamiltonians (including electronic structure Hamiltonians (5.8), the Ising model, Heisenberg model etc.) it can be truncated to a number of terms which is polynomial in the size of the system. The basic intuition for this fact is that arbitrarily strong, arbitrarily long-range interactions do not appear in nature.

Calculations in quantum chemistry are generally concerned with the energy $E = \langle \hat{H} \rangle = \langle \psi | \hat{H} | \psi \rangle$ of a state $|\psi\rangle$ under the Hamiltonian $\hat{H}$. By linearity, this is given by

$$E = \langle \psi | \hat{H} | \psi \rangle = \sum_{i\alpha} h^i_a \langle \psi | \hat{\sigma}^i_a | \psi \rangle + \sum_{ijab} h^{ij}_{ab} \langle \psi | \hat{\sigma}^i_a \otimes \hat{\sigma}^j_b | \psi \rangle + \cdots \qquad (5.36)$$

Thus the energy reduces to a weighted sum over a polynomial number of expectation values of local Pauli observables, and can be precisely estimated by means of repeated local single-qubit measurements together with classical floating-point addition. For an N-qubit state, we can thus efficiently evaluate the expectation value of a $2^N \times 2^N$ Hamiltonian.

In classical methods for quantum chemistry, as we have already seen, the ground state energy of the chemical Hamiltonian is generally found by an iterative variational method, in which a nonlinear optimization algorithm is used to minimize the energy with respect to the parameters $|\phi\rangle$ of a scalable ansatz for the state. The electronic configuration of the molecule is approximately represented in the digital logic of the central processing unit (CPU) by means of an approximate, scalable ansatz $f(\vec{\phi}) = |\psi(\vec{\phi})\rangle$. This restricts the CPU to a small subspace of $\mathscr{H}$. The ansatz parameters $\vec{\phi}$ are initialized according to a guess or approximate method, and the energy $\langle \psi(\vec{\phi}) | \hat{H} | \psi(\vec{\phi}) \rangle$ is evaluated by a numerical method. The optimization algorithm then attempts to iteratively drive towards the ground state.

We propose a hybrid quantum-classical analogue to this approach, illustrated schematically in Fig. 5.4. Rather than a CPU, we make use of a small quantum processor (QPU), constructed from the universal gate set, to implement the ansatz and evaluate the energy of candidate ground states. The QPU takes as input some real parameters $\vec{\phi}$, and prepares a state $|\vec{\phi}\rangle$ in a qubit register of the device. Copies of this

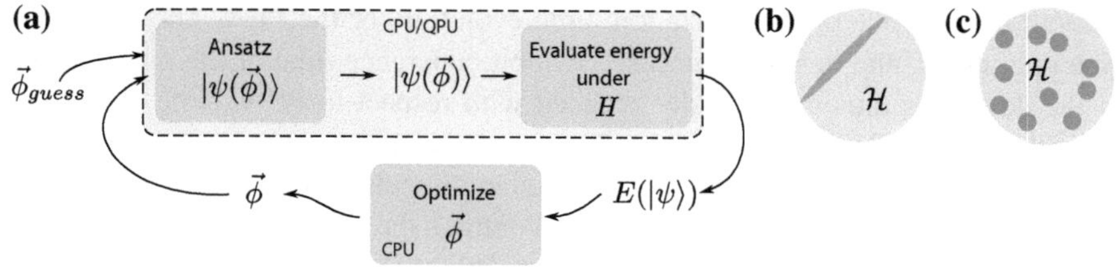

Fig. 5.4 Quantum simulation without time-evolution. (**a**) Classical approaches to quantum chemistry often make use of the variational method. An approximate ansatz is chosen for $|\psi\rangle$, allowing a subspace of $\mathscr{H}$ to be represented in the CPU. The ansatz parameters $\vec{\phi}$ are initialized according to some approximate method, and a nonlinear optimization algorithm then iteratively minimizes the energy of the state under the chemical Hamiltonian. We propose a hybrid quantum-classical analog to this approach, in which a small quantum processor (QPU), likely constructed from a universal gate set, is used in place of the CPU to implement the ansatz and compute energies (*red box*) while the optimization algorithm still runs on a classical processor (*blue box*). (**b, c**) Various classical ansätze exist to efficiently parametrize small subspaces of the electronic configuration Hilbert space. A QPU cannot scalably address the full Hilbert space, but should nonetheless give access classically intractable ansätze (color online)

state are then measured in a number of local Pauli bases—corresponding to terms in (5.36)—from which the energy is recovered by classical floating-point addition. The optimization process, which updates the ansatz parameters $\vec{\phi}$ based on the current energy, is then performed classically on the CPU.

We have already seen that existing efficient classical ansätze are limited to describing certain classes of chemical systems. By representing the trial wavefunction on a quantum device, although we can still only parametrize a polynomial subspace of $\mathscr{H}$, the expectation is that we should nonetheless be able to efficiently implement *different class* of ansätze for which no efficient classical algorithm exists. In particular, there is reason to believe that efficient parametrization of highly correlated, entangled electronic configurations should be more efficient using a QPU than a CPU. A simplistic argument for the existence of such ansätze is as follows: Consider a quantum circuit, parametrized by a number of classical control phases and constructed somehow at random from a universal gate set. It is then reasonable to believe that the chance of finding an efficient classical parametrization of the output state $|\psi(\vec{\phi})\rangle$ should be vanishingly small. Hence it is likely that there exist a large number of classically intractable ansätze which can be implemented using exponentially fewer resources on a QPU. The development of such ansätze remains an open problem in quantum simulation.

5.5.2 Advantages

Quantum chemistry using the PEA promises full-configuration-interaction calculations using relatively few qubits but requires an imposing number of gate operations, due in part to the Trotterization overhead required for time evolution. Our approach, although limited to an approximate ansatz, provides variationally optimal solutions without dependence on Trotterization, time-evolution, or the PEA. The number of gate operations, and hence the necessary qubit coherence time or the physical size of the device, is thus dramatically reduced with respect to PEA. Note that in our algorithm, the QPU repeatedly prepares a state under the ansatz and immediately measures it in a local basis, destroying all quantum coherence—this is the entirety of the "quantum" stage of computation. In contrast, the PEA must remain coherent throughout. A recent numerical investigation [26] into FCI-PEA computation of the ground state energy of iron sulfide Fe_2S_2 estimated the required calculation time—which, for a ballistic computation, is equivalent to the required coherence time—to be 1.5 years.

By implementing a large fraction of the total computation on a classical processor, we ensure that the use of quantum resources is limited to the operations where they give the greatest advantage, i.e. in the representation of quantum states. The trade-off with respect to the PEA is that we no longer have ballistic, single-shot computation, since the classical optimization algorithm must make a large number of calls to the QPU before convergence to the ground state is achieved.

5.5.3 Scaling

A single call to an n-qubit QPU prepares $|\psi(\vec{\phi})\rangle$ and returns the expectation value of a tensor product of Pauli operators. The gate cost of the state preparation stage is dictated by our choice of ansatz, which is not predetermined—we *assume* that we will always choose an ansatz with a known decomposition into a polynomial number of gate operations, without explicitly defining this choice.

The measurement stage can be parallelized, giving an estimate of a single term in (5.36) with precision p in after $O(|h|^2/p^2)$ repeated measurements of copies of the state. This leads to a total readout cost $O(|h_{max}|^2 M/p^2)$ to evaluate the energy of a trial state $|\psi\rangle$ under the full Hamiltonian, where M is the number of terms in the expansion (5.36).

5.5.4 Open Questions

In an ideal world, having chosen a classically intractable class of chemical systems of interest, we would then design a QPU which addresses the subspace of $\mathscr{H}$ in which they live. While PEA-based methods provide an explicit prescription for the necessary gate-model circuit, much less is known when it comes to the deterministic design of circuits which efficiently parametrize the approximate ansätze required for our algorithm. This currently limits the scope of our method, and restricts our ability to assess its asymptotic performance.

Further uncertainty arises as a result of the use of nonlinear numerical optimization. How many calls to the QPU will it take for the classical minimization to traverse the quantum energy landscape and converge to the ground state? How will the choice of optimization algorithm affect the precision in E? These questions depend in turn on the choice of ansatz and the nature of the chemical Hamiltonian. In our experimental demonstration (Sect. 5.6) we use a general-purpose optimization algorithm (Nelder-Mead simplex, `fminsearch` in Matlab / `scipy.fmin` in Python). Despite experimental imperfections this algorithm performed well on a realistic chemical Hamiltonian, converging to the ground state to acceptable accuracy after a few hundred iterations. There is scope for considerable optimization in the choice of this optimization algorithm, and we expect that existing techniques from "classical" quantum chemistry should be directly applicable to our scheme. This is not to say that the optimization will *always* run in polynomial time, and thus the scalability of our approach remains an open question.

An interesting problem for all quantum simulation algorithms is raised by the intractability of full quantum state tomography. Although our algorithm and PEA-based methods both prepare the approximate ground state, from which the ground-state energy can be efficiently obtained, we cannot recover full information on the eigenstate vector—in order to do so we would need an efficient classical parametrization of the state, which presumably does not exist for those chemical problems which demand quantum simulation. Similarly, we cannot obtain the full spectrum of the Hamiltonian, since in general it has exponentially many eigenvalues. Therefore, through quantum simulation we can at best hope to obtain partial information on the eigenstates of classically intractable Hamiltonians, for instance measuring expectation values of some operator of interest other than the Hamiltonian, or partial information on the spatial configuration of a molecule. The design and optimization of such readout methods will be an important problem for future implementations of our algorithm.

5.6 Experiment

We have performed a proof-of principle experimental demonstration of this method, using the CNOT-MZ device previously described. Since the CNOT-MZ permits arbitrary 2-qubit state preparation as well as arbitrary measurement in a local Pauli basis, it is an ideal test-bed for our algorithm. The fact that the chip is fully computer-controlled allows the optimization feedback loop to be performed without human intervention, which is important when a single run of the experiment can involve thousands of unique measurement settings.

The ability to prepare two-qubit states (Sect. 2.2.7) allows us to investigate 4×4 Hamiltonians. It is interesting to draw comparison with the recent experimental demonstration by Lanyon et al. [29], in which two photonic qubits were implemented in a bulk optical setup. In this work the authors make use of a more orthodox PEA-based algorithm and as such are forced to use one qubit as the control register which leaves room for a 2×2 Hamiltonian only.

We chose the helium hydride ion HeH^+ as the chemical system of interest for this demonstration. Helium hydride is the strongest known acid, and was likely the first molecule to form in the universe after the big bang. The second-quantized Hamiltonian for the two-electron system of He-H^+ can be expressed as a 4×4 matrix using a minimal atomic basis set (STO-3G). The coefficients $h^i_\alpha \ldots$ in the expansion of the Hamiltonian were calculated by means of an FCI method in the PSI3 ab-initio computational chemistry package [32]. Note that this approach is not scalable in general, and is used here for convenience only. The mapping from qubits to fermions is performed using the Jordan-Wigner transform, as described in Sect. 5.4.1.

In our experimental implementation, owing to the small size of the circuit used, we choose as an ansatz the full-two qubit Hilbert space, which has 6 free parameters. This provides a robust test for the performance of the optimization algorithm, but is not at all scalable. Future demonstrations will need to implement a scalable ansatz, the design of which remains an open problem.

Figure 5.5a shows experimental data from a typical optimization run, with the energy converging to the ground state after ~100 iterations of the algorithm. We studied the degree of entanglement of the two-qubit state as a function of time during the optimization run, using as a metric the *tangle* $T = \mathcal{C}^2$, where $\mathcal{C}$ is concurrence (1.40). For the case of HeH^+ we found that while the algorithm does pass through

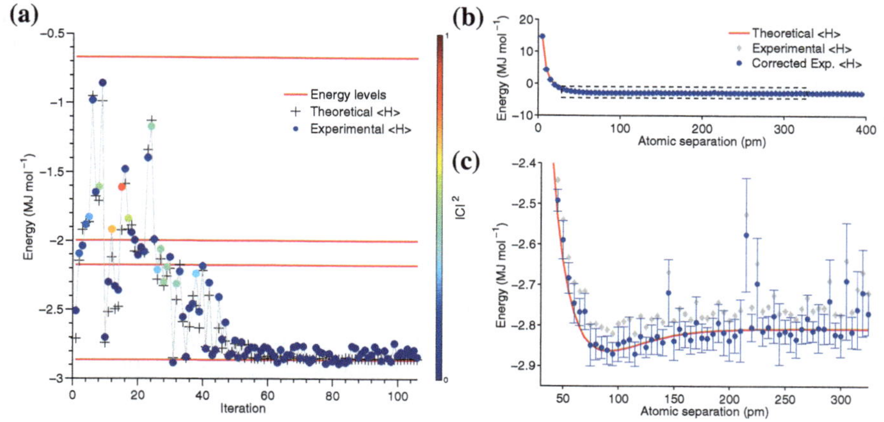

Fig. 5.5 **a** A single optimization run, finding the ground state energy of HeH^+ for a specific molecular separation, $R = 90\,\text{pm}$. *Coloured points* show the experimentally computed energy as a function of the optimization step, where the colour corresponds to the tangle of the 2-qubit state, estimated directly from $|\phi\rangle$. *Red lines* show the four eigenenergies of the FCI Hamiltonian of HeH^+ in a minimal basis. *Crosses* correspond to a theoretical ideal value of the energy, computed at each optimization step. **b** Experimentally measured bond dissociation curve of HeH^+, analogous to the approximate Lennard-Jones potential. Each point corresponds to the ground state energy of the Hamiltonian $\hat{H}(R)$ for a particular value of the atomic separation R, and is obtained from a single optimization run as shown in (**a**). The *red line* shows the theoretical *curve*, and *grey points* show experimental data prior to correction for a small systematic error. **c** is a magnified region of (**b**), demonstrating that our experimental setup can resolve the dip in the curve, corresponding to the equilibrium bond length of the molecule

regions of strongly entangled Hilbert space during the optimization run, the qubit representation of the final electronic ground state was generally only very weakly entangled. The nature of the Jordan-Wigner transform is such that there is not necessarily a correspondence between the degree of entanglement of the fermionic state and that of its qubit representation.

Writing the HeH^+ Hamiltonian as a function of the atomic separation R,

$$\hat{H} = \sum_{i\alpha} h^i_a(R)\,\hat{\sigma}^i_a + \sum_{ijab} h^{ij}_{ab}(R)\,\hat{\sigma}^i_a \otimes \hat{\sigma}^j_b + \cdots \tag{5.37}$$

we repeated the optimization process for several values of R, thus obtaining the bond dissociation curve shown in Fig. 5.5b, c. This curve is analogous to the Lennard-Jones potential previously described. The equilibrium bond length—the atomic separation of the molecule in its relaxed state—was measured to be $R = 92.3 \pm 0.1$ pm, with a corresponding ground-state electronic energy of $E = -2.865 \pm 0.008$ MJ/mol. Here the error bar is due only to Poissonian finite statistics, and does not take into account error introduced by other experimental imperfections or the convergence of the optimization algorithm. The experimental data in Fig. 5.5 correspond to tens of thousands of unique measurements on two-photon states generated by the CNOT-MZ, and as such represent the most demanding application of the device to date.

5.7 Discussion

In this work we have simulated the HeH^+ molecule using a single two-qubit gate together with a handful of single-qubit rotations. By comparison, the PEA-based FCI method for an equivalent Hamiltonian, without adiabatic state preparation, would require at least 12 CNOT operations. By dispensing with the need for PEA, Trotterization and time evolution, our algorithm enables much more chemistry to be done with much less quantum hardware, and dramatically reduces the necessary coherence time. In doing so, however, we introduce new unknowns. In particular, it is not clear whether the optimization algorithm will necessarily converge in polynomial time. Furthermore, we have not provided a deterministic technique by which a given polynomially-sized ansatz may be parametrized in terms of a quantum circuit—a fundamental requirement for both theoretical analysis and practical implementation of our algorithm.

5.8 Statement of Work

My main contribution in this section was in the optimization and maintenance of the CMOZ-MZ chip, together with theoretical analysis of the work. Figure 5.5 is due to Alberto Peruzzo, who also measured the data.

References

1. S. Michael, D. Auld, C. Klumpp, A. Jadhav, W. Zheng, N. Thorne, C.P. Austin, J. Inglese, A. Simeonov, A robotic platform for quantitative high-throughput screening. Assay. Drug. Dev. Technol. **5**, 637–657 (2008)
2. R.P. Feynman, Simulating physics with computers. Int. J. Theor. Phy. Theor. Phy. **21**, 467–488 (1982)
3. P.W. Shor, *Polynomial-Time Algorithms for Prime Factorization and Discrete Logarithms on a Quantum Computer* (1995). arXiv:quant-ph/9508027
4. R. Babbush, A. Perdomo-Ortiz, B. O'Gorman, W. Macready, A. Aspuru-Guzik, *Construction of Energy Functions for Lattice Heteropolymer Models: A Case Study in Constraint Satisfaction Programming and Adiabatic Quantum Optimization* (2012). arXiv:1211.3422
5. G.J. Halász, A. Perveaux, B. Lasorne, M.A. Robb, F. Gatti, Á. Vibók, Simulation of laser-induced quantum dynamics of the electronic and nuclear motion in the ozone molecule on the attosecond time scale. Phys. Rev. A **86**, 043–426 (2012)

6. P.W. Anderson, The resonating valence bond state in la2cuo4 and superconductivity. Science **235**(4793), 1196–1198 (1987)
7. R. Moessner, S.L. Sondhi, P. Chandra, Two-dimensional periodic frustrated ising models in a transverse field. Phys. Rev. Lett. **84**, 4457–4460 (2000). May
8. J.W. Britton, B.C. Sawyer, A.C. Keith, C.-C.J. Wang, J.K. Freericks, H. Uys, M.J. Biercuk, J.J. Bollinger, Engineered two-dimensional Ising interactions in a trapped-ion quantum simulator with hundreds of spins. Nature **484**, 489–492 (2012). April
9. M. Sarovar, A. Ishizaki, G.R. Fleming, K.B. Whaley, Quantum entanglement in photosynthetic light-harvesting complexes. Nat. Phys. **6**, 462–467 (2010). June
10. Dirac. *Proceedings of the Royal Society of London. Series A, Containing Papers of a Mathematical and Physical Character*, vol. 123, 714–733 (1929)
11. L. Pauling, *The nature of the chemical bond*, vol. 2 (Univ. Press, New York, addison-wesley edition, Cornell, 1939)
12. M. Born, J. Robert Oppenheimer, On the quantum theory of molecules. Ann. Phys. **389**, 457–484 (1927)
13. I. Kassal, J.D. Whitfield, A. Perdomo-Ortiz, M.-H. Yung, A. Aspuru-Guzik, Simulating chemistry using quantum computers. Ann. Rev. Phys. Chem. **62**, 185–207 (2011)
14. J.C. Slater, The theory of complex spectra. Phys. Rev. **34**, 1293–1322 (1929)
15. I. Shavitt, R.J Bartlett, *Many-Body Methods in Chemistry and Physics: MBPT and Coupled-Cluster Theory*. Cambridge (2009)
16. W. Koch, M.C. Holthausen, *A Chemist's Guide to Density Functional Theory, 2nd Edition*. Wiley (2001)
17. T. Fukuhara, P. Schauß, M. Endres, S. Hild, M. Cheneau, I. Bloch, C. Gross, *MicroscopiC Observation Of Magnon Bound States And Their Dynamics* (2013). arXiv:1305.6598
18. M.A Nielsen, The fermionic canonical commutation relations and the jordan-wigner transform (2005)
19. P. Jordan, E. Wigner, über das paulische äquivalenzverbot. Z. Phys. **47**(9–10), 631–651 (1928)
20. G. Ortiz, J.E. Gubernatis, E. Knill, R. Laflamme, Quantum algorithms for fermionic simulations. Phys. Rev. A **64**, 022319 (2001)
21. R. Somma, G. Ortiz, J.E. Gubernatis, E. Knill, R. Laflamme, Simulating physical phenomena by quantum networks. Phys. Rev. A **65**(4), 042323 (2002)
22. D.S. Abrams, S. Lloyd, Simulation of many-body fermi systems on a universal quantum computer. Phys. Rev. Lett. **79**, 2586–2589 (1997)
23. A.Y. Kitaev, A.H. Shen, M.N. Vyalyi, *Classical and Quantum Computation*. Amer Mathematical Society (2002)
24. Michael A. Nielsen and Isaac L. Chuang. *Quantum Computation and Quantum Information (Cambridge Series on Information and the Natural Sciences)*. Cambridge University Press, 1 edition (2004)
25. J. Whitfield, J. Biamonte, A. Aspuru-Guzik, Simulation of electronic structure Hamiltonians using quantum computers. Mol. Phys. **109**, 735–750 (2011). March
26. D. Wecker, B. Bauer, B.K. Clark, M.B. Hastings, M. Troyer, Can quantum chemistry be performed on a small quantum computer? (2013). arXiv:1312.1695
27. O. Johnston, *The Illusion of Life: Disney Animation*. Disney Editions (1981)
28. Alán Aspuru-Guzik, Anthony D. Dutoi, Peter J. Love, Martin Head-Gordon, Simulated quantum computation of molecular energies. Science **309**(5741), 1704–1707 (2005)
29. B.P. Lanyon, J.D. Whitfield, G.G. Gillett, M.E. Goggin, M.P. Almeida, I. Kassal, J.D. Biamonte, M. Mohseni, B.J. Powell, M. Barbieri, A. Aspuru-Guzik, A.G. White, Towards quantum chemistry on a quantum computer. Nat. Chem. **2**, 106–111 (2010)
30. S. Lloyd, Universal quantum simulators. Science **273**, 1073–1078 (1996)
31. J. Kempe, A. Kitaev, O. Regev, *The Complexity of the Local Hamiltonian Problem*. arXiv:quant-ph/0406180
32. http://www.psicode.org/

Chapter 6
Increased Complexity

> *Actually, if we wanted to, although it's expensive, we could put detectors all over [. . .] and build up the whole curve simultaneously. . .*
>
> Feynman

6.1 Introduction

A survey of the literature reveals a rich history of experiments in which p photons are sent through an optical circuit with m modes. The experimentalist looks to see where the photons went, examining spatio-temporal correlations using an array of single-photon detectors, in an effort to determine whether the experiment is (i) working properly and/or (ii) doing anything interesting.

In general, the number of possible detection patterns across m modes grows as $\binom{m}{p}$, and can be extremely large even for modest values of m and p. It is therefore often convenient or even essential to use a greater number of detectors $d \leq m$ than photons, allowing $\binom{d}{p}$ patterns to be monitored simultaneously. True number-resolving single photon detectors (Sect. 1.6.4) are not currently widely available. However, number-resolving detection can instead be achieved using multiplexed non-number-resolving detectors, again demanding the ability to operate and monitor many detectors in parallel.

In this chapter we describe experiments using $\leq$5 photons in $\leq$21 modes, leading to tens of thousands of possible events. In order to efficiently assess the physics and performance of these experiments we need a detection system akin to a camera, capable of recording and correlating events across many detectors in parallel. To this end, we have constructed a novel detection system using 16 Si APD single photon detectors, supported by electronics and hardware capable of simultaneously monitoring all possible p-fold detection events up to $p = 16$ in real-time. Owing to the capacity of this machine to efficiently photograph quantum states with very large

P. Shadbolt, *Complexity and Control in Quantum Photonics*,
Springer Theses, DOI 10.1007/978-3-319-21518-1_6

Hilbert space dimension, we euphemistically refer to it as a *Hilbert space telescope*. This system has so far enabled at least three experiments which otherwise would not have been possible, two of which are described in this section.[1]

In Sect. 6.3 we describe experiments using up to five photons in structured and unstructured interferometers, designed to implement both *quantum walks* and the so-called BosonSampling problem. We reconstruct time-correlated images of the multiphoton output state of these devices, observing a clear signature of generalized bosonic bunching in Hilbert spaces of up to ∼50,000 dimensions. We make use of this capability to test two unique approaches to efficient verification of BosonSampling.

6.2 Time-Correlated Single Photon Counting

In multiphoton experiments we are frequently presented with the problem of measuring correlation functions in space or time, based on detection events over d detectors. Very often, this problem reduces to the counting of *coincidences*. By postselecting on events in which p detectors fired within some small coincidence time-window Δt, we record only those events in which all p photons were generated in the same down-conversion event or femtosecond pulse, preserving temporal indistinguishability and thus high-visibility quantum interference.

Certain experiments require more precise timing information. For example, the pulse envelope of a laser or the delay introduced by a coaxial cable can be measured using the closely related techniques of temporal *autocorrelation* and *cross-correlation*. Here, the exact time of each detection event is measured and recorded with very high (fs) precision by a fast clock. The recorded arrival time of a single detection event is referred to as a *time-tag*. The autocorrelation function $G(\tau)$ of continuous time-varying signal $I(t)$

$$G(\tau) = \lim_{T\to\infty} \frac{1}{2T} \int_{-T}^{T} I(t) \cdot I(t+\tau) dt \tag{6.1}$$

provides information about similarity of the signal with a delayed version of itself, while the cross-correlation function between two signals $I_{1,2}(t)$

$$G_{12}(\tau) = \lim_{T\to\infty} \frac{1}{2T} \int_{-T}^{T} I_1(t) \cdot I_2(t+\tau) dt \tag{6.2}$$

measures the similarity between these signals as a function of the delay between them. When counting discrete photon detection events with finite timing resolution, the signal is no longer analog and we instead compute the discretized quantities

[1] The third, completed very recently, is described in a pre-print [1] due to Matthews et al.

$$G(t) = \sum_t N(t) \cdot N(t+\tau)\ ; \quad G_{12}(\tau) = \sum_t N_1(t) \cdot N_2(t+\tau) \tag{6.3}$$

where $N_i(t)$ is the number of photons detected in channel i and timebin t. These functions can be easily computed from measured timetags.

6.2.1 TCSPC Hardware

In all of the multi-photon experiments described here, silicon-based avalanche photodiode single photon detectors (APDs) have been used for detection. A number of alternative single photon detector technologies are described in detail in Sect. 1.6.4 of this thesis. Upon absorbing a single photon within the frequency band to which the APD is sensitive (~500 to ~900 nm), a 3.3 V trigger/timing logic (TTL) voltage pulse is generated with finite probability—typically ~60% for the Perkin-Elmer APDs used here.

The task of TCSPC is then to count and correlate these TTL pulses in time. The rate of single-photon detection is typically on the order of MHz, and many standard data acquisition systems do not have sufficient bandwidth, channel count, or timing resolution to capture and correlate events at this rate. As a result, TCPSC largely depends on dedicated high-speed electronics falling into one of two categories. pure coincidence-counting systems, often based on nuclear instrumentation module (NIM) logic or field programmable gate arrays (FPGAs), and time-to-digital converters, which convert incoming TTL pulses into high-resolution digital timetags to be processed downstream.

The counting systems used in Chaps. 2, 3, 4 and 5 of this thesis are custom-built around Xilinx Virtex 5 or Spartan 6 FPGAs. These devices provide a lithographically-fabricated array of $\sim 10^5$ logic units (gates), which can be reconfigured to implement dedicated coincidence counting logic with much greater bandwidth and timing stability than can be achieved using general-purpose ICs. These systems are built to count instances of $O(10)$ possible coincidence patterns over ≤ 8 independent channels, with a fixed coincidence window of 5 ns, and do not provide high-resolution timing information—instead simply reporting the number of events for each pattern over a fixed integration time of 1 s.

6.2.2 DPC-230

The DPC-230 is a 16-channel photon correlator, produced by Becker and Hickl GmbH. It is principally designed for multiphoton fluorescence spectroscopy and biological imaging, however we have adapted it for applications in quantum photonics. The principal functionality of interest is the capability of the DPC-230 to time-tag incoming TTL pulses on 16 independent channels simultaneously with ~80 ps resolution, allowing coincidence counting using an array of 16 Si APDs. The instrument, which is packaged as a PCI card for installation in a standard PC, uses 16 CMOS time-to-digital converters (TDCs) to record the absolute arrival time of TTL pulses.

The design and interface of the DPC-230 are focussed on multiphoton spectroscopy and biological imaging, and the device is largely configured for off-line analysis of small samples—a few seconds of photon time-tag data. It is not intended for real-time use, and does not provide coincidence counting as built-in functionality. For instance, all time-tag data must be written to a hard disk and post-processed before it can be used, and all of the documentation and bundled software are written with this mode of operation in mind. However, in the context of quantum photonics the experimentalist needs both real-time operation, providing immediate feedback when working in the lab, as well as the ability to integrate for days or weeks at a time in multiphoton experiments where the n-fold detection rate is extremely low. We therefore built a custom hardware/software stack which addresses these issues, providing coincidence counting functionality and allowing the DPC-230 to be operated in realtime, accumulating up to ten million photon time-tags per second, for months at a time.

The internal architecture of the DPC-230, together with the custom PC hardware/software stack, is shown in Fig. 6.1. Two TDC chips, each having 8 independent

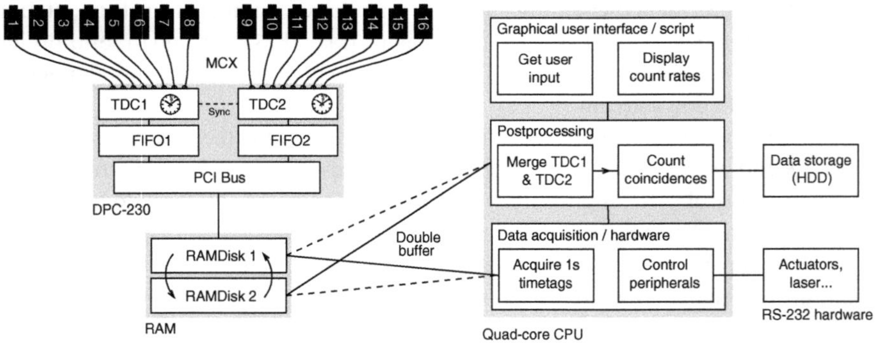

Fig. 6.1 Two groups of eight Perkin-Elmer Si APDs send TTL pulses via MCX coaxial cable to the DPC-230 TCSPC (time-tagging) board. The DPC-230 uses two 8-channel TDC chips, which time-tag the rising edge of incoming pulses with ps resolution. These timetags are stored in one of two FIFO buffers, each of which can store 2 million photons at a time. Coincidence counting and control is managed by three processes running in parallel on a quad-core desktop computer. The first, highest-priority process sequences time-tagging and periodically reads timetags from the PCI bus into one of two RAMDisks, operating in a double-buffered arrangement. This process also communicates via RS-232 with other high-priority hardware such as the Ti:Sapphire laser and SMC100 motor controllers. While this process is acquiring timetags, coincidence counting is performed on older data in a parallel process. Optimized C code merges data from TDC1 and TDC2 and then counts/stores all 2^{16} possible N-fold coincidence events, up to $N = 16$, with a variable top-hat coincidence window (typically 5 ns). This stage also implements 16 arbitrary software-defined delays, allowing path length differences and APD idiosyncrasies to be accounted for. Finally the count rates are filtered and summed according to the users request, and plotted in a real-time GUI

TDC channels, are synchronized by a stable clock. These convert TTL pulses generated by single photon detectors into 24-bit timetags, encoding the channel number and absolute time of arrival of each pulse, down to a bin width of 82.3 ps. These timetags are temporarily stored in first-in-first-out (FIFO) buffers, each of which is capable of storing $\sim 2 \times 10^6$ photons. This data is read into RAM in the host PC over a standard PCI bus.

At high photon count rates (up to 1×10^7 photons/s), around 30 MB of timetag data is acquired per second. Since for multiphoton experiments we must often continuously integrate for a number of days, it is essential that this data is processed in real-time so that unmanageably large volumes of timetags do not accumulate. In order to achieve maximum throughput we use two high-priority processes, written in optimized C and running on separate cores of a Pentium Core i7 CPU to implement data acquisition and post-processing/coincidence counting in parallel. Timetags are acquired to the DPC-230's internal FIFOs for one second, and are then read into one of two RAMDisk buffers by the data-acquisition process. This data is passed to the post-processing thread, which merges data from the two TDC chips and then counts and stores instances of all possible p-fold coincidences up to $p = 16$, with a user-specified coincidence window. Above a net detection rate of $\sim 1 \times 10^6$ photons per second, this process takes slightly longer than one second to process one-second's worth of timetag data. The data acquisition thread must therefore wait for the post-processing thread to "catch up", resulting in a reduced duty cycle and a linear decrease in the effective n-fold detection efficiency. The system is routinely used at a throughput of $\sim 5 \times 10^6$ photons/s (Fig. 6.2).

In order to avoid storing integrated count-rates for all 2^{16} possible events, we exploit the sparsity of the data—fivefold events and above are very rate—and write

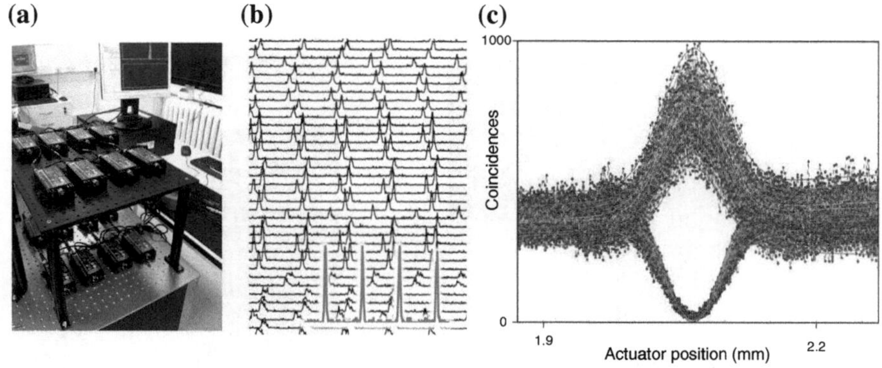

Fig. 6.2 The sheer amount of information generated by the DPC-230 demands new approaches to data processing and analysis. **a** An array of 16 Perkin-Elmer Si APDs. **b** 36 cross-correlation curves, acquired in a single 2-s measurement. The *red curve* shows a typical cross-correlation function. The time between peaks corresponds to the repetition rate of the Ti:Saph, i.e. $\sim$12.5 ns. Detector jitter is the predominant source of broadening of the peaks, giving a FWHM of $\sim$1 ns. **c** 105 Hong-Ou-Mandel dips measured in parallel over a single actuator scan, using a type-II pulsed SPDC source and the DPC-230 (colour online)

only nonzero countrates to disk. Despite the significant saving in disk space provided by this sparse format, it was necessary to further optimize the representation of post-processed data by means of a custom binary file format, which stores coincidence data together with information pertaining to the motor controllers, laser, and other metadata. This file format is described in detail in Appendix A. Finally, this data is sent to a graphical user interface, running in a third process, where it can be graphed, filtered and analysed by the experimentalist.

6.2.2.1 User Interface

A bad craftsman blames her tools, but correlation is not causation—we may not infer that a person who blames their tools is unskilled. With the rapid increase in the complexity of experimental apparatus and the volume of data generated by tools such as the DPC-230, we must take greater care over the interface between the human being and their experimental setup. When actively developing and optimizing apparatus in the lab, the importance of responsive control and immediate, intuitive feedback cannot be understated.

We have built a graphical user interface (GUI), shown in Fig. 6.3a, which enables experimental control, real-time analysis, post-processing, and management of coincidence data from the DPC-230. This GUI also interfaces with SMC100 motor controllers and the Coherent *Chameleon* Ti:Sapphire laser. The user can choose an arbitrary subset of detection events of interest, including number-resolved patterns under a variety of pseudo-number-resolving schemes, to be displayed and graphed in real-time. This interface also controls the integration time, coincidence window and software delays, and allows arbitrary sequences of measurement and automation to be scripted.

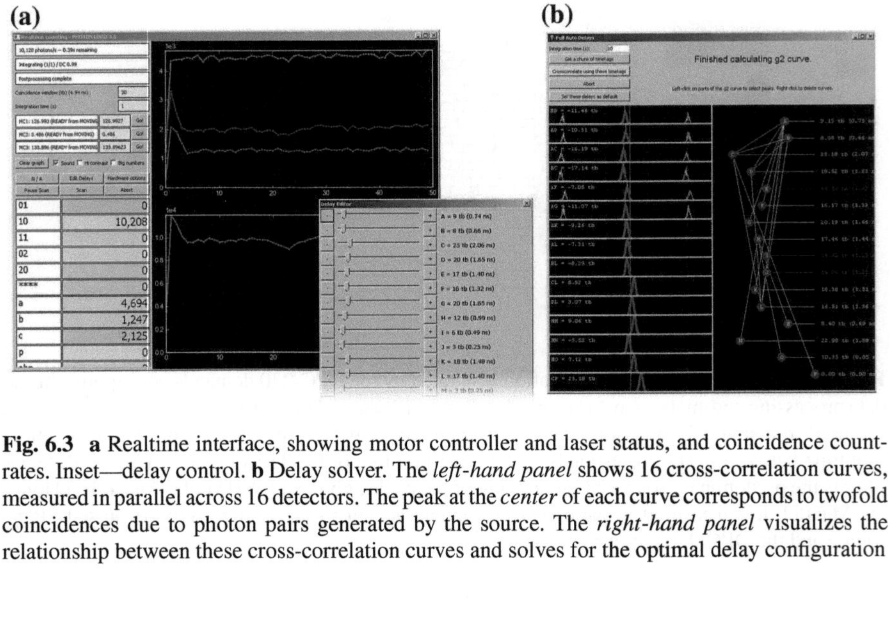

Fig. 6.3 **a** Realtime interface, showing motor controller and laser status, and coincidence countrates. Inset—delay control. **b** Delay solver. The *left-hand panel* shows 16 cross-correlation curves, measured in parallel across 16 detectors. The peak at the *center* of each curve corresponds to twofold coincidences due to photon pairs generated by the source. The *right-hand panel* visualizes the relationship between these cross-correlation curves and solves for the optimal delay configuration

6.2.2.2 Delays

Synchronization of delays is an important consideration when coincidence-counting with large numbers of detectors. For example, digital pulse-conditioning logic inside the Perkin-Elmer APD assembly, together with variation in cable and free-space path lengths, can introduce up to $\sim$20 ns of delay between detection of a photon and arrival of the corresponding TTL pulse at the TDC. We must therefore introduce artificial delays into "early" channels, ensuring that timetags due to photons generated within the same downconversion event or laser pulse fall within the coincidence window of the counting logic. Traditionally, this has been accomplished using rack-mounted delay boxes, which simply switch between fixed lengths of coaxial cable. The optimization of these delays has typically been performed by a process of trial and error on behalf of the experimentalist. With many more detectors to deal with, this optimization process becomes very time consuming.

These issues can be mitigated by making direct use of timing information provided by the DPC-230. First, physical electronic delay boxes are no longer required, since all delays can be implemented in software—simply by shifting timetags from each channel by some user-specified time Δt. Secondly, the task of finding optimal delay configurations has been almost entirely automated. Switching out of coincidence-counting mode, we acquire timetags for $\sim$10 s and then compute cross-correlation functions (6.2) between all possible pairwise combinations of channels. These G_{12} curves are analyzed by a physically-inspired optimization process which automatically finds the optimal delay configuration with minimal input from the user—see Fig. 6.3b. This capability has been essential for the multiphoton experiments described in Sect. 6.3, where frequent changes in the detection setup required regular re-calibration of delays.

6.3 Multiphoton Quantum Interference

All experimental work in this thesis has so far been performed in a qubit encoding. Although we have studied two photons in up to 6 modes—a system with 21 unique configurations—we have only been interested the four two-qubit states $|00\rangle \ldots |11\rangle$, and we have postselected on detection events which fall in that subspace. This has allowed us to directly exploit the majority of the established language and theory of quantum computation, much of which is written in terms of qubits i.e. in the circuit model. In particular, we have made use of proofs of universality and scaling such as those of KLM (1.6.2) to guide our experimental design. The fact that the literature should be so focussed on qubits is not surprising—as with a classical computer, any finite d-dimensional *qudit* encoding can be efficiently and exactly represented in terms of two-level systems, which often have advantages in terms of simplicity of analysis and hardware efficiency.[2]

At present, the major bottlenecks for the development of universal LOQC are the lack of deterministic, scalable sources of indistinguishable photons, and the difficulty of optical-frequency adaptive measurements. Although work is under way to develop deterministic sources in a variety of architectures (see Sect. 1.6.3), current technology is very much "pre-threshold", and experiments which go beyond four photons remain challenging. On the other hand, with the advent of integrated quantum photonics, *modes* are comparatively cheap—reconfigurable silicon photonic devices with hundreds of waveguides are readily available [4].

Perhaps we can obtain a greater computational return per photon, at least in the short term, by dispensing with the circuit model and making direct use of a larger number of optical modes? Taking a simple example, if our basic resource is 5 photons, then using the $2p$ modes that are minimally required to encode p independent qubits, we generate a Hilbert space on qubits of dimension $2^5 = 32$. If on the other hand we inject the same 5 photons into a device with 25 modes, we generate a Hilbert space with dimension 118,775. Naïvely, we might expect that it is in general hard to classically compute the effects of quantum interference in such scenarios. Moreover, it is not obvious that this computational advantage should depend on adaptive measurements, and the associated problem of GHz feed-forward, required for universal LOQC. The price we pay for this experimental convenience is the guarantee of universality provided by KLM and others—see Sect. 1.5.4—but it is nonetheless conceivable that we might retain an exponential quantum speedup for specific tasks.

As well as an alternative approach to photonic quantum computation, this section also introduces a new attitude towards quantum interference. Even when studying fundamental physical phenomena such as entanglement and nonlocality (for example in Chaps. 4 and 5), we have so far treated photonic quantum interference as a *resource* which powers the CNOT-P gate, rather than a basic physical phenomenon of interest. In this section we will demonstrate complex multiphoton quantum interference effects which have not previously been observed and are of basic scientific interest in their own right, irrespective of potential practical applications.

Large-scale experiments of this form can currently only be implemented in a controlled way using the technology previously described: first, the ability to build intrinsically stable multi-path interferometers on a monolithic chip, and second, a detection system capable of efficiently acquiring a detailed picture of the full output state. The theoretical framework described in Sect. 1.5.3 will also be indispensable for the numerical simulation and verification of experimental results.

The variety of possible linear optical networks which can be constructed from beamsplitters and phase-shifters is infinite. Here we will consider two extrema: the *most structured* and *least unstructured* nontrivial interferometers. In the first case,

[2]In a circuit model architecture, replacing qubits with d-level systems has been shown to give a modest multiplicative $\log_2 d$ advantage in the number of gate operations [2] and facilitate controlled-$\hat{U}$ operations [3].

we construct linear, symmetric arrays of uniformly coupled waveguides. Using these devices, we implement *quantum walks* of up to five photons, which continuously tunnel back and forth between neighbouring waveguides in the array. In the unstructured case, we use *Haar random* circuits, chosen uniformly at random from the space of all possible interferometers. A recent result by Aaronson and Arkhipov [5] has shown that multiphoton experiments using randomized interferometers of this type are very likely to be classically intractable, even without feed-forward. We experimentally test various aspects of this scheme, referred to as BOSONSAMPLING, using up to 3 photons. Finally we discuss the problem of verification and validation of BOSONSAMPLING machines, and experimentally demonstrate the potential utility of quantum walks in this context.

6.3.1 Quantum Random Walks

6.3.1.1 Galton's Board

A *Galton board* is constructed by hammering nails into a board so as to form a regular lattice, as shown in Fig. 6.4. The board is mounted vertically, and a ball is dropped from above. Upon striking each pin the ball bounces at random, either to the left or the right. Each row of the lattice corresponds to a discrete timestep, and we are usually only interested in the ball's lattice site k on the current row, rather than its exact position in space. The ball is said to take a *random walk* through the lattice, and is referred to as a *walker*. Random-walk dynamics appear throughout nature, from Brownian motion and neuroscience to the hunting tactics of sharks [6] and humans [7]. Moreover, random walks form the basis for a number of randomized classical algorithms, including graph connectivity [8] and machine learning.[3] Interestingly, the best-known approximate polynomial-time algorithm for the permanent (see Sect. 1.5.3) of a nonnegative real matrix, due to Jerrum et al. [10], makes use of a random walk. A random walk is a *Markov chain*, as the instantaneous stochastic dynamics of the walker do not depend on the past trajectory or history.[4]

The time evolution of the ball in a Galton board is discretized. At a given timestep, corresponding to one row of the board, the walker bounces to a neighbouring lattice site—either to the left or the right, with equal probability. After n_t timesteps, there are $\binom{n_t}{k}$ possible routes that the walker might have taken to arrive at a site k. The probability that the walker arrives at the kth site is therefore binomially distributed,

$$P(k) = \frac{1}{2^n}\binom{n_t}{k} \tag{6.4}$$

[3] In fact, a classical random walk was used as part of a machine learning algorithm to optimize the performance of the CNOT-MZ chip [9].

[4] Note that the momentum of the ball in the Galton board gives the system some memory of past states, and the system is therefore only approximately Markovian.

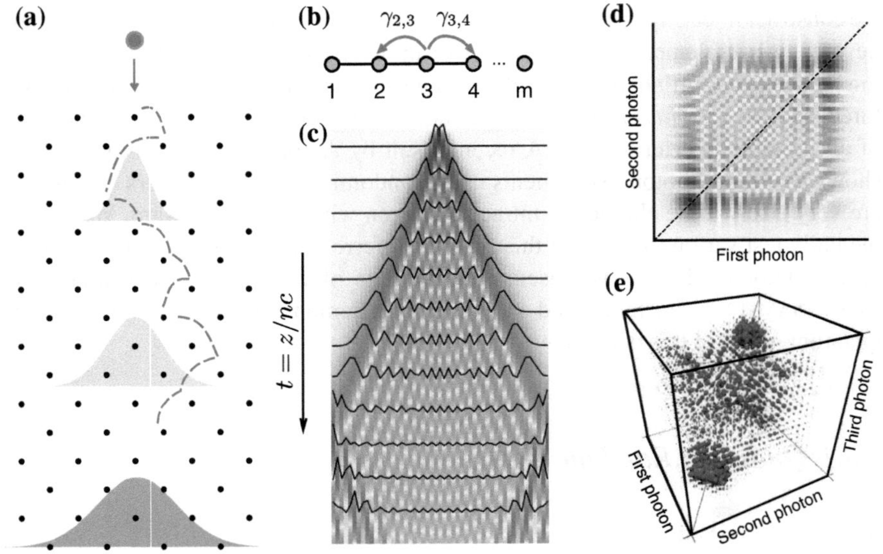

Fig. 6.4 Classical and quantum random walks. **a** A ball takes a classical random walk through the pins of a Galton board. The probability that the ball lands at a given lattice site is binomially distributed. **b** The state space of a walker can be represented as a graph, whose vertices and edges correspond to lattice sites and allowed trajectories of the walker, respectively. **c** A single quantum walker injected into a discrete array of continuously-coupled lattice sites undergoes a quantum walk, continuously tunnelling to neighbouring sites. The wavefunction spreads ballistically, and interferes with itself to create wavelike patterns in the probability distribution. This numerical simulation also shows reflection of the wavefunction at the edge of the lattice. **d** The dynamics of a single walker can be reproduced classically, for instance using water waves. However, if two indistinguishable walkers are simultaneously injected into adjacent modes we obtain quantum interference, leading to generalized bosonic bunching which has no classical analog. Photon pairs are more likely to be detected at nearby sites, i.e. on the main diagonal of the correlation matrix. **e** Injecting three photons into adjacent sites, we observe the higher-order equivalent of (**d**), where photons are again clustered on the main diagonal. In general, the correlation matrix of p photons can be represented as a p-dimensional hypercube

where the centre of the distribution corresponds to the starting site k_0. This behaviour is shown in Fig. 6.4a.

6.3.1.2 Quantum Walks

At any given time, a classical random walker occupies a single site k in the lattice. What happens if we instead use a quantum walker, able to occupy a coherent superposition state $|\psi\rangle = \sum_{k=1}^{m} d_k |k\rangle$ over many lattice sites k? There are many ways to construct such quantum-mechanical analogues of Galton's board, all of which fall under the banner of *quantum walks*. All quantum walks have in common the fact that the walker is a quantum particle, and that the stochastic evolution is described by a lossless (unitary) process. Most quantum walks are characterised by a time-independent or periodic Hamiltonian with a regular, local, graph-like structure.

A number of basic phenomena distinguish quantum walks of a single particle from classical random walks. First, the probability distribution over sites, an example of which is shown in Fig. 6.4c, is qualitatively more complex than that of classical particles, owing to interference of the wavefunction with itself. This interference pattern often features two prominent *ballistic lobes* of high probability, whose distance from the origin is a linear function of the evolution time. A quantum walker thus traverses the lattice faster than a classical particle, in the sense that after fixed amount of time we are more likely to detect the quantum particle at a greater distance from the origin.

Quantum walks provide a generic, simple model of quantum dynamics, and as such have found a broad range of practical applications. Quantum walks have been used to model natural quantum phenomena including photosynthesis [11] and exciton dynamics [12], and form the basis of a variety of quantum algorithms for problems including search [13–15], verification of matrix products [16], evaluation of balanced binary game trees [17], and computation of a broad class of general formulas [18]. Moreover, quantum walks have been shown to provide a basic primitive for universal quantum computation [19, 20]—an idea which can be traced back to Feynman, who describes a computer with a time-independent Hamiltonian in *Simulating physics with computers* [21].

6.3.1.3 Continuous-Time Quantum Walks of Photons

All quantum walks can be categorized as being either *continuous time* or *discrete-time*. The discrete-time quantum walk [22–24] is perhaps the closest quantum ancestor of a Galton board. The system evolves in discrete timesteps, during which the evolution of the state of the walker is described by a fixed unitary operator $\hat{W}$. By analogy with Galton's nail, $\hat{W}$ places the walker into a coherent superposition of leftward and rightward motion, resulting in a superposition (usually balanced) over lattice sites at the next timestep. The time evolution of a discrete quantum walk is then generated by repeated application of $\hat{W}$, with $|\psi\rangle_{\text{out}} = \hat{W}^{n_t}|\psi\rangle_{\text{in}}$ after n_t timesteps.

In this work we are instead concerned with *continuous-time* quantum walks [25, 26], which do not share such a strong analogy with Galton's board. Discrete and continuous-time quantum walks have been shown to be equivalent in the limit of an infinitely small timestep [27]. Rather than using instantaneous splitting operations, a continuous-time walk creates a constant opportunity for a walker at a particular site to leak or tunnel into some subset of other sites in the lattice. This opportunity, or *coupling*, has an associated strength which is related to the rate at which probability amplitude moves between a particular pair of connected sites.

More formally: the state of a single walker in a lattice with m sites $k = \{1, 2 \dots m\}$ can always be written in a basis $\{|1\rangle, |2\rangle \dots |m\rangle\}$

$$|\psi\rangle(t) = \sum_{k=1}^{m} b_k |k\rangle = \sum_{k=1}^{m} b_k \hat{a}_k^\dagger |\mathbf{0}\rangle \tag{6.5}$$

where $|k\rangle$ is the state of a walker in the kth site, with a corresponding creation operator $\hat{a}_k^\dagger$. Following Childs et al. [25], the connectivity of the lattice can be represented as a graph G, whose vertices and edges correspond to lattice sites and site-to-site couplings respectively. For the simple example of the Galton board, G is a 1-D linear graph with nearest-neighbour couplings, as shown in Fig. 6.4b. Any G—and therefore any lattice—can be written as an $m \times m$ *generator* matrix M, where an element M_{ij} corresponds to the coupling strength between sites i and j of the lattice.

To see the physical meaning of these couplings, we first examine a classical continuous-time random walk. Let $P_i(t)$ be the probability of finding the walker at site i and time t. If two sites i, j, are coupled with a strength M_{ij}, then it is reasonable to think that the rate of change of probability at a site should be proportional to both the coupling strength and the probability distribution over all adjacent sites:

$$\frac{dP_i(t)}{dt} = \sum_{j=1}^{m} M_{ij} P_j(t). \tag{6.6}$$

To a first approximation, this reproduces the results of Galton's board—in particular, since P_i are positive real numbers, no interference effects are seen.

For quantum states, time evolution is governed by the Heisenberg equation (1.20), and a quantum walker prepared at site i evolves according to

$$i\frac{d\hat{a}_i^\dagger(t)}{dt} = \left[\hat{a}_i^\dagger(t), \hat{H}\right]. \tag{6.7}$$

We can then model analogous dynamics to (6.6) for a single quantum walker by choosing a Hamiltonian in the interaction picture

$$\hat{H} = \sum_{i,j=1}^{m} M_{ij}\hat{a}_i^\dagger \hat{a}_j. \tag{6.8}$$

where we have set $\hbar = 1$, leading to

$$i\frac{d\hat{a}_i^\dagger(t)}{dt} = -\sum_{j=1}^{m} M_{ij}\hat{a}_j^\dagger(t). \tag{6.9}$$

Terms on the diagonal of M can be interpreted as coupling a site to itself, encouraging the walker to stay at a particular site.

Integrated photonics provides a particularly simple route to the implementation of continuous-time quantum walks. Arrays of straight, parallel, evanescently coupled waveguides can be lithographically fabricated in a variety of material systems, providing a compact, interferometrically stable lattice upon which a walker, in the form of coherent laser light or single photons, can move. Each waveguide then represents a site in the lattice, and the time parameter corresponds to longitudinal distance z in the array, with $t = z/(nc)$ where n is the refractive index of the material. The $M_{i \neq j}$ correspond to the strength of evanescent coupling between adjacent pairs of waveguides, which can be precisely controlled as described in Sect. 2.2.2. Since the evanescent field of a single mode waveguide falls off exponentially with distance (Sect. 1.5.1), the coupling strength between next nearest-neighbour waveguides is exponentially weaker than that of nearest-neighbours, and can usually be neglected.

A number of experiments report the use of laser-written waveguides in 3-D architectures to implement walks on highly-connected graphs [28, 29]. Moreover, the Reck-Zeilinger scheme described in Sect. 1.5.4 of this thesis allows graphs with any connectivity to be experimentally implemented in a 2-D waveguide structure. However the majority of implementations, including those reported in this thesis, use a 2-D nearest-neighbour array, leading to a 1-D lattice such as that shown in Fig. 6.4b. Each site is then coupled to at most two nearest neighbours, giving a simple tridiagonal form for the generator:

$$M_{ij} = \begin{cases} \gamma_{ij}, & \text{if } |i-j| = 1\,, \\ \beta_i, & \text{if } i = j, \\ 0, & \text{otherwise.} \end{cases} \rightarrow \begin{bmatrix} \beta_1 & \gamma_{12} & 0 & 0 & 0 & 0 & 0 \\ \gamma_{12} & \beta_2 & \gamma_{23} & 0 & 0 & 0 & 0 \\ 0 & \gamma_{23} & \beta_3 & \gamma_{34} & 0 & 0 & 0 \\ & & & \cdots & & & \\ 0 & 0 & 0 & 0 & 0 & \gamma_{m-1,m} & \beta_m \end{bmatrix}, \quad (6.10)$$

where γ_{ij} are evanescent couplings with $\gamma_{ij} = \gamma_{ji}$ and β_i are waveguide propagation constants. The Hamiltonian for a single particle on a 1-D lattice is then

$$\hat{H} = \sum_{j=1}^{N} \beta_j \hat{a}_j^\dagger \hat{a}_j + \gamma_{(j,j-1)} \hat{a}_{j-1}^\dagger \hat{a}_j + \gamma_{(j,j+1)} \hat{a}_{j+1}^\dagger \hat{a}_{j1}. \quad (6.11)$$

A waveguide array with a fixed length z is then described by an $m \times m$ unitary operator $\hat{U}$, which acts on the single-particle Hilbert space $\mathscr{H}_m^1$ and is equivalent to the transfer matrix Λ

$$\Lambda \leftrightarrow \hat{U} = e^{-i\hat{H}z/nc}\,; \quad |\psi\rangle_{\text{out}} = \hat{U}|\psi\rangle_{\text{in}} \quad (6.12)$$

which under the assumption of zero loss completely characterises the device.

Quantum walks of a single particle have now been reported in a variety of physical systems including cold atoms [30], ions [31, 32], and nuclear magnetic resonance [33], as well as a large number of optical implementations [34–36]. Single-photon quantum walks have been used to simulate the band structure of strained graphene [28] and the relationship between decoherence and the quantum/classical boundary [37].

In optical single-particle quantum walks, the walker is either implemented using a single photon or a single beam of coherent laser light. In the absence of two-photon quantum interference, the dynamics are thus described by a classical wave theory (see Sect. 1.5) and both classical and quantum light sources give identical detection probabilities. In other words, the interference pattern in Fig. 6.4c can be exactly reproduced using water waves. This implies that these experiments and associated algorithms can be simulated on a digital computer with an overhead at most polynomial in the system size [38]. Thus no quantum algorithm based on a single-particle quantum walk provides any more than a polynomial (likely quadratic [22]) speedup over a classical computer, and all such algorithms can be simulated with a constant $O(1)$ scaling using classical wave computers. The only exceptions to this rule are oracle-based algorithms, for example the result of [39].

In order to see quantum walk behaviour which is not explained by a classical wave model, we must introduce multiple contiguous walkers to the lattice [40]. The first experimental demonstration [41] used photon pairs generated by SPDC together with an array of 21 uniformly coupled SiO_xN_y waveguides. This work has since been extended, using entangled photons to simulate fermionic statistics [42], as well as observation of two-photon time evolution [29]. A number of recent demonstrations have used laser-written waveguides to implement discrete-time walks of two to three indistinguishable photons [43–45], including walks in 3-D structures [46]. Quantum walks of two interacting magnons have also recently been observed, using cold atoms trapped in a linear lattice [47].

Let's consider a 1-D array of uniformly coupled waveguides, such as that used by Peruzzo et al. [41]. Measuring the twofold coincidence count-rate between single-photon detectors at output ports i and j, we can plot a correlation matrix (Fig. 6.4d), showing the probability of coincidental detection of photons in any given pair of waveguides (i,j). We find that indistinguishable photons are very likely to be detected either at the same site, or at adjacent waveguides. Specifically, we observe two clouds of probability density, centred about the main diagonal $(i = j)$ of the correlation matrix, corresponding to events in which both photons are detected in the same half of the array. Events in which the photon pair is split across the array (off-diagonal terms in the correlation matrix) are suppressed.

This effect is a generalized form of two-photon quantum interference (Sect. 1.5.3), and has no classical analogue. For the case of $m = p = 2$, for example, we recover exactly the situation of Hong, Ou and Mandel. Intuitively, this effect can be thought of as a consequence of the known tendency of photons to bunch together. It should be noted that in a two-photon quantum walk, in contrast with HOM interference at a 50:50 BS, it is not always the case that both photons are detected at exactly the same site. The observed increase in probability of coincidental detection at *nearby but not identical* sites will be referred to here as *clouding*, to distinguish from HBT-style *bunching*—it is not clear that the two are equivalent.

In order to calculate states and probabilities in multi-particle walks, we make direct use of the method outlined in Sect. 1.5.3. To re-iterate, any p-photon amplitude or probability can be expressed as the permanent of a $p \times p$ submatrix of the $m \times m$ transfer matrix Λ. As we have already seen, Λ is equivalent to the single-particle unitary time evolution operator $\hat{U}$ (6.12), which is a direct function of the single-particle Hamiltonian (6.11). This is the origin of the nomenclature of M as a generator, since it is a relatively small $m \times m$ matrix operating on $\mathscr{H}_m^1$ which is used to *generate* $\hat{H}$ and $\hat{U}$ on the much larger multi-particle Hilbert space $\mathscr{H}_m^p$, which has dimension $\binom{m+p-1}{p}$.

How hard is it to simulate such highly-ordered quantum walks of many indistinguishable photons? Looking at figures such as those shown in Eq. 6.4, we might expect that by exploiting the apparent structure of the probability distribution we should be able to efficiently predict the outcome of such experiments using a classical algorithm. Indeed, there have been tentative theoretical efforts to approximately model such distributions in terms of Bessel functions [48]. However, our best known exact methods depend on the calculating the permanent, which in general is exponentially hard. Detailed discussion of these issues, in a slightly different context, is given in the next section.

6.3.2 BosonSampling

Imagine a computer which can be built in the real world. The ECT says that any such apparatus can be efficiently simulated by a probabilistic Turing machine — "Time on all reasonable machine models is related by a polynomial." [49]. While the standard Church-Turing thesis (Sect. 2.1) has been all but proven for practical purposes [50], the ECT has been significantly weakened by the prospect of quantum computing. The idea that some machines might be fundamentally classically intractable is uncomfortable, and the veracity of the ECT remains the subject of intense debate [51]. It is reasonable that this debate should be serious: before making any large financial investment in quantum computing research, we should first ensure that the problems of interest cannot be efficiently solved using classical computers.[5]

Shor's algorithm constitutes the best-known challenge to the ECT. A machine running Shor's algorithm could not currently be simulated in polynomial time. However, Shor's algorithm does not yet render the ECT untenable, as it has not been proven that factoring is classically intractable, i.e. outside P. Moreover, *even if* Factoring $\not\subset$ P, it may still be the case that undiscovered new physics or decoherence phenomena render the construction of a scalable quantum factoring machine fundamentally (as opposed to practically) impossible (see, for example [52]).

Can we find *experimental* evidence against the ECT? It would arguably be very convincing if we could build a universal, fault-tolerant quantum computer capable of significantly out-performing a classical computer at problems such as prime factoring. Although great progress has been made, the largest number factored so far using a quantum computer is 21 [53].[6] In contrast, the current recommended RSA key length (i.e. the size in bits of a composite number whose prime factorization is

[5] The ECT is not sufficiently well-posed to ever be formally disproved, only weakened.

[6] or 143, depending on how you define "quantum computer" [54].

considered classically intractable in the near-term) is $L = 2048$. To run Shor's algorithm on this key would require $O(L^2)$ logical qubits, which after error-correction would likely correspond to billions of coherent components [55]. Even if we *could* build such a machine,[7] we would *also* need to prove that factoring is hard in order to strike a blow against the ECT.

As discussed in Chap. 5, exact simulation of quantum chemistry and superconducting materials is currently classically intractable. Non-universal quantum simulation, which is likely to be technologically less demanding than universal quantum computing [56], is believed to provide an exponential quantum speedup in some instances and would represent a challenge to the ECT. However, this approach suffers from the same burden of proof as Shor's algorithm: it is even harder to formally *prove* that such problems are classically intractable.

BOSONSAMPLING, proposed in 2010 by Aaronson and Arkhipov [5], attempts to solve the problems of theoretical proof and experimental difficulty outlined above. Although BOSONSAMPLING holds for any noninteracting boson, for simplicity we will only consider photons. We can then define the problem:

> Build an m-mode interferometer A, whose transfer matrix Λ is chosen uniformly at random from the space of all possible interferometers (i.e. by the Haar measure, Sect. 1.3.1). Place a detector at the output of each mode, and inject $p \lesssim \sqrt{m}$ indistinguishable photons to different input ports of the circuit. BOSONSAMPLING is the problem of generating a single detection event, sampled from the probability distribution $\mathcal{B}$ over all possible p-fold detection events.

The device, consisting of an interferometer together with p photons and m detectors, is referred to as a *boson computer*. We have already shown in Sect. 1.5.3 of this thesis that each element of the probability distribution $\mathcal{B}$ can be computed as a permanent of a $p \times p$ submatrix of Λ. The core result of [5] is to show that, given certain very reasonable conjectures, fast approximate classical algorithms for BOSONSAMPLING would have very dramatic and unlikely consequences for existing models of computation:

> Suppose there exists a classical algorithm which takes as input a description of a boson computer A and an error bound ε, and samples from an approximate distribution $\mathcal{B}'$ such that $||\mathcal{B} - \mathcal{B}'|| \leq \varepsilon$, in poly($|\Lambda|$, $1/\varepsilon$) time. Then $GPE_{\times}$, which is a #P-hard problem, is solvable in $\mathsf{BPP}^{\mathsf{NP}}$. [5]

Here $GPE_{\times}$ is the problem of estimating the permanent of a matrix of complex Gaussian random numbers $X \sim \mathcal{N}_{\mathbb{C}}^{p \times p}$ to multiplicative $\pm\varepsilon \cdot p!$ error, with high probability. In 1979 it was proven by Valiant [57] that calculation of the permanent is #P-complete, and an exact fast algorithm would imply $\mathsf{P} = \mathsf{NP}$. Here, #P is the class of problems which *count* the solutions of decision problems in NP. Polynomial-time approximate randomized algorithms for the permanent of certain classes of matrix exist—for example those due to Jerrum et al. [10] (real, positive matrices) and Gurvitz [58] (complex matrices with atypically large permanents), but no known algorithm achieves the generality, precision and success probability demanded by

[7] A CPU containing in excess of a billion nanoscale transistors can be bought for less than £10.

BosonSampling. Much of the work of [5] is to provide evidence that $GPE_\times$—i.e. approximate estimation of the permanent of a random complex matrix—is #P-hard, and to prove that if so, a fast classical BosonSampling machine would imply $\mathsf{P}^{\#\mathsf{P}} = \mathsf{BPP}^{\mathsf{NP}}$, collapsing the polynomial hierarchy (P, NP, coNP etc.) to an extent that would have far-reaching implications, not least rendering postselected[8] classical computers as powerful as postselected quantum computers ($\mathsf{BPP}_{\mathsf{path}} = \mathsf{PostBQP}$).

Arguably, BosonSampling provides even stronger evidence against the ECT than Shor's algorithm. If Factoring turns out to be in P, although existing public-key cryptography would be broken, we would not have to modify our existing models of computation. If on the other hand BosonSampling has an efficient classical algorithm, then a generic, foundational assumption of basic computation complexity theory would fall.

At the same time, in practical terms BosonSampling is a *weaker* than Shor's result. Factoring is a problem with known real-world applications, while BosonSampling does not have any such known use. It is important to emphasize that a BosonSampling machine does not allow one to compute the permanent, only to *sample* from $\mathcal{B}$. A necessary condition for the proof is that $m \gtrsim p^2$, in which case each element of $\mathcal{B}$ is exponentially small in p. We therefore cannot simply run the machine many times in order to well-estimate a particular entry in $\mathcal{B}$.

From an experimental point of view, the most compelling feature of BosonSampling is the relative ease with which an advantage over existing classical machines can be achieved. Rapid scaling in p, together with a number of experimentally convenient properties, renders BosonSampling a leading candidate for the first experimental quantum speedup over classical computers. Specifically:

- The exponential difficulty of classical BosonSampling scales *particularly fast*. In numerical simulations, using an optimized implementation of the fastest known exact algorithm[9] for the permanent [59], we find that for $p > 6$ (with $m = p^2$) full calculation of $\mathcal{B}$ becomes practically impossible on a single GHz CPU. It is reasonable to think that experiments with $\gtrsim$20 indistinguishable photons in $\gtrsim$400 modes will begin to challenge even for existing supercomputers. 8-photon experiments have been reported using photons generated by SPDC [60].
- BosonSampling depends on high-visibility quantum interference, but, in contrast with KLM, does not require adaptive measurement or feed-forward—techniques which, at optical frequencies, remain experimentally very challenging.
- Given the fidelity with which linear-optical networks and single-photon detectors can be constructed, it is not necessarily the case that BosonSampling machines require error correction (Sect. 1.4.1), although this remains an open question (see, for example, Ref. [61]).

[8]Allowing postselection on exponentially unlikely outcomes for both quantum and classical machines.

[9]Benchmarks and optimized Cython code for the permanent are given in Appendix A.

Let's assume that we have a BOSONSAMPLING machine with $p > 20$. How can we verify that the machine is truly implementing BOSONSAMPLING, and that experimental imperfection has not caused it to output a classically tractable distribution? The success or failure of Shor's algorithm can be easily checked in polynomial time by simply multiplying the prime factors. All problems in NP have this promise, however, the output of problems in $P^{\#}\mathsf{P}$ cannot necessarily be checked in polynomial time. Indeed, the original proposal of BOSONSAMPLING suggests that efficient verification might be *fundamentally* impossible.

An intuitive argument was recently given by Gogolin et al. [62], who consider the problem of distinguishing a BOSONSAMPLING machine, which samples from $\mathcal{B}$, from a fake, classical uniform-sampler, which samples p-fold clicks from the flat distribution $\mathcal{F} : P_i = 1/d \, \forall \, i$. Since Λ is Haar-random, $\mathcal{B}$ is roughly uniform. Moreover, when $m \gtrsim p^2$, $\mathcal{B}$ is spread roughly uniformly over exponentially many possible detection patterns. It might therefore appear that $\mathcal{B}$ should be well-approximated by $\mathcal{F}$. Indeed, the authors show that without knowledge of Λ, the experimentalist would need to obtain an exponential number of samples from a machine under test before they could distinguish $\mathcal{B}$—generated by a "real" BOSONSAMPLING machine—from $\mathcal{F}$. If the purpose of BOSONSAMPLING is to provide experimental evidence against the ECT, this is a serious problem.

Previous experimental implementations of BOSONSAMPLING have used up to four indistinguishable photons, together with randomized interferometers constructed using optical fibre [63], lithographically fabricated waveguide chips [64] and laser-written waveguides [65, 66]. These early demonstrations have largely focussed on verification of the relationship between measured statistics and permanents of Λ. In our experimental work we have instead attempted to address the more recent questions of verification and validation of BOSONSAMPLING, including the potential role of quantum walks in this problem.

6.3.3 Experiment

We have performed three and four-photon experiments, using photonic chips to implement both quantum walks (QW) and Haar-random BOSONSAMPLING unitaries (RU). Throughout our experimental work, we have focussed on characterisation of the bosonic clouding effects described in Sect. 6.3, the problems of BOSONSAMPLING verification outlined in Sect. 6.3.2, and the potential relationship between the two.

A full schematic of the experimental setup is shown in Fig. 6.5. A multi-photon type-I SPDC source (Sect. 6.3.3) is coupled into PMF fibre. These fibres are butt-coupled to the input ports of either the QW (Sect. 6.3.3) or RU (Sect. 6.3.3) chip. Photons are then coupled out of the chip and detected/correlated using the counting system previously described, together with an array of fibre splitters for pseudo-number resolving detection (Sect. 6.3.3).

6.3.3.1 Multiphoton Source

The photon source used in this experiment, illustrated in Figs. 6.5 and 6.6a, is based on type-I down-conversion (Sect. 1.6.3) in the pulsed regime. A Ti:Sapphire pulsed laser (Coherent *Chameleon Ultra II*) generates 144 fs FWHM pulses at 780 nm, with a repetition rate of 80 MHz. The average output power is ∼3.7 W, with peak power in excess of 300 kW. This light is attenuated using a zero-order HWP together with a

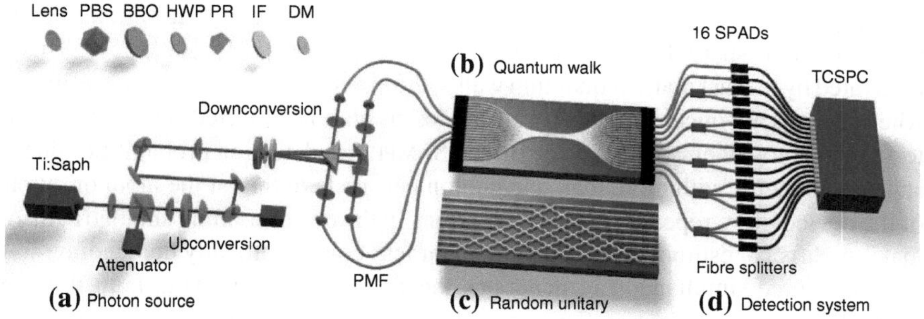

Fig. 6.5 Experimental setup to generate (**a**), interfere (**b**, **c**) and detect (**d**) single photons. **a** 780 nm laser light from a 140 fs pulsed Titanium:Sapphire laser was attenuated with a *HWP* and a *PBS*, before frequency doubling with a type-I *BBO* nonlinear crystal. The subsequent 390 nm light was reflected from four *DMs* and focused onto a type-I BiBO nonlinear crystal to generate double pairs of photons through spontaneous parametric down conversion. After passing through an *IF*, photons are reflected off a prism (*PR*) and collected into polarisation maintaining fibres which are butt-coupled, via a V-groove fibre array, to either (**b**) the QW chip, or (**c**) the RU chip. Outgoing photons are coupled from the chip using a second fibre array, either directly to 16 APD detectors (**d**), or via a network of fibre splitters. Detection events are time-correlated and counted using a 16-channel TCSPC

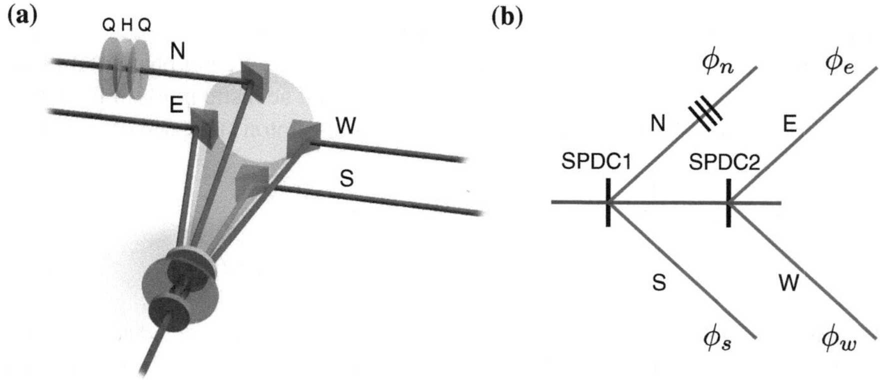

Fig. 6.6 **a** By coupling to the four compass points of the *SPDC* cone (*N*, *S*, *E*, *W*) we can well-approximate states of three degenerate photons in three modes, heralded on detection of a fourth photon. **b** This approach can be modelled as two independent *SPDC* processes at different crystals

Glan Taylor high-power PBS, and is upconverted to 390 nm using a 2 mm-thick BBO, phase-matched for colinear SHG. This pump beam is cleaned of 780 nm light using four DMs and is then focussed to a waist of ~40 μm on a 2 mm-thick BiBO crystal phase-matched for type-I downconversion, generating photon pairs at 780 nm on a cone with 3° opening angle. The pump is then removed using a DM together with a band-pass IF (Semrock *MaxLine*, $\lambda_0 = 780$ nm, $\Delta\lambda = 3$ nm, transmission ~95 %). Four prisms, together with precision alignment stages and aspheric lenses, are used to couple downconverted light from the four compass points of the downconversion cone (0° N, 90° E, 180° S, 270° W) into PMF.

Our goal is to use this source to generate states of three and four photons where no two photons share the same mode—i.e. the Fock states $|111\rangle$ and $|1111\rangle$. This is motivated by the fact that, for quantum walks, the observed dynamics are more diverse when photons are injected into separate modes, as photons injected at the same mode tend to stick together. Similarly, for BosonSampling, detection probabilities due to state components with more than one photon per mode (either at the input or output of the device) have repeated rows and columns in the corresponding submatrix of Λ, making classical estimation of the permanent less computationally demanding.

This source simultaneously generates the SPDC state (1.168) at both the N/S and E/W compass points of the cone. This can be modelled as two independent downconversion processes, and is more easily visualized as simultaneous SPDC at two independent crystals as shown in Fig. 6.6b. Assuming the filters, collection optics, and geometry are symmetric across all four modes, we can write the output state as

$$\begin{aligned} |\Psi\rangle &= |\psi\rangle_{SPDC}^{ns} \otimes |\psi\rangle_{SPDC}^{ew} \qquad (6.13) \\ &= [|0_n 0_s\rangle + e^{i(\phi_n+\phi_s)}\gamma|1_n 1_s\rangle + e^{2i(\phi_n+\phi_s)}\gamma^2|2_n 2_s\rangle] \\ &\quad \otimes [|0_n 0_s\rangle + e^{i(\phi_e+\phi_w)}\gamma|1_e 1_w\rangle + e^{2i(\phi_e+\phi_w)}\gamma^2|2_e 2_w\rangle] + \text{h.c.} \qquad (6.14) \end{aligned}$$

where the phases $\vec{\phi}$ arise due to differences in path length between the each collection stage and the BiBO crystal. These free-space optical paths are not phase-stabilized, and therefore fluctuate randomly with temperature and acoustic noise in the lab. In this work we are principally concerned with the four-photon subspace of (6.14),

$$\begin{aligned} |\Psi(\vec{\phi})\rangle_4 = \frac{1}{\sqrt{3}}\big[&e^{i(\phi_n+\phi_s+\phi_e+\phi_w)}|1_n 1_s 1_e 1_w\rangle \\ &+ e^{2i(\phi_e+\phi_w)}|0_n 0_s 2_e 2_w\rangle + e^{2i(\phi_n+\phi_s)}|2_n 2_s 0_e 0_w\rangle\big]. \qquad (6.15) \end{aligned}$$

With high pump power there is a chance that one photon will be detected in each of the four modes, in which case the state is projected onto $|1111\rangle$ term only, and the global phase can be ignored. However, if the modes are mixed by an interferometer prior to measurement, we can no longer be sure that a given fourfold detection event did not arise from one of the $|2200\rangle$ or $|0022\rangle$ terms. Since the phases $\vec{\phi}$ fluctuate in time, the average state will in general be partially mixed, and will not produce high-visibility quantum interference.

We take a number of measures to overcome these problems. First, we can easily perform three-photon experiments in which modes N, S, E are sent into the interferometer, while mode W is connected directly to a heralding single-photon detector. Postselection on detection of three photons at the output of the interferometer together with detection at the herald then isolates an effective input state of three degenerate single photons in three modes, $|111\rangle$.

In order to study four-photon statistics arising from the $|1111\rangle$ term, we must currently take a less satisfactory approach. Connecting all four modes to the interferometer, we first acquire fourfold coincidence countrates c_i^m using the full four-mode four-photon SPDC state (6.14). During this measurement, we continuously rotate the polarization of one arm of the source using an arrangement of waveplates (Fig. 6.6), forcing the average state into a maximal mixture[10]

$$\begin{aligned}\hat{\rho}_4 &= \int (\hat{R}(t)_n \otimes I_s \otimes I_e \otimes I_w)|\Psi(t)\rangle_4\langle\Psi(t)|_4 dt \\ &= |1111\rangle\langle 1111| + |2200\rangle\langle 2200| + |0022\rangle\langle 0022|. \end{aligned} \tag{6.16}$$

We would then like to treat detection events due to the $|2200\rangle$ and $|0022\rangle$ terms as noise. Fortunately, these countrates can be experimentally measured: making two further measurements with modes E/W and N/S disconnected from the interferometer respectively, we obtain two new sets of experimental countrates (c_i^{ns}, c_i^{ew}). Subtracting these countrates from the mixed state data c_i^m, we recover statistics which model the behaviour of the desired $|1111\rangle$ state. This approach is problematic, quantum interference is to a certain extent artificially constructed using measurements on a maximally mixed state. As a result, this is not a scalable route to high photon-number experiments. However, short of post-selecting from higher-photon number terms in the state with exponentially low probability, or waiting for a scalable single-photon source, it nonetheless provides an immediate route to experimental tests of the $|1111\rangle$ state.

6.3.3.2 Quantum Walk Chip

All of the quantum walk data presented in this section was measured using a 2-D waveguide array (Fig. 6.5), fabricated in silicon oxynitride (SiO_xN_y). The coupled region of the array, which is 700 µm long, consists of 21 waveguides with a cross-section of 2.2 µm × 0.85 µm, and a uniform pitch of 1.3 µm. Curved fan-in and fan-out waveguides connect each mode to input and output ports at the chip facets, which

[10] By introducing a strong, controlled, uniform source of noise, we "override" any effects from the uncontrolled, non-uniform thermal/acoustic phase fluctuation. In this sense, the method described here shares some similarity with the techniques for precise characterization under environmental noise described in Sect. 4.5.

are butt-coupled to PMF V-groove arrays with a pitch of 127 μm. The waveguides are tapered to a width of 0.7 μm to improve coupling to the fibre mode. An oil-based index-matching fluid was used to further improve coupling efficiency. The lumped fibre-to-fibre coupling was typically ∼30 %.

SiO_xN_y is a ceramic material, whose refractive index can be tuned between ∼1.45 and ∼2 by controlling the nitrogen/oxygen ratio (x/y). Compared to silica-on-silicon waveguides (Sect. 2.2.1), SiO_xN_y can achieve a much higher refractive index contrast of $\Delta = (n_2^2 - n_1^2)/2n_1^2 = 4.4\,\%$ between the waveguide core (n_2) and cladding (n_1), allowing a significantly smaller bend radius (Sect. 1.5.1) and more compact fan-in/fan-out regions.

6.3.3.3 BosonSampling Chip

Following the prescription of Aaronson and Arkhipov [5], the BosonSampling device used here implements a random unitary operation on 9 modes, chosen by the Haar measure (Sect. 1.3.1) on $U(9)$. In order to implement this operator in linear optics, we make use of the Reck-Zeilinger scheme described in Sect. 1.5.4 of this thesis. The layout of directional couplers is shown in Fig. 6.5.

The chip is fabricated in silicon nitride (Si_2N_3), with a refractive index contrast of 27 %. The device consists of a total of 36 directional couplers. The high index-contrast afforded by Si_2N_3 was essential in order to achieve a compact circuit and suppress losses. Each waveguide has a cross-sectional width of 1.5 μm, and the pitch between parallel waveguides was designed to match that of the fibre arrays (127 μm). At each directional coupler, the separation between waveguides is 2.5 μm, with an interaction length, depending on the desired coupling ratio, of ∼400 μm. This device is not reconfigurable—each coupling ratio and internal phaseshift was written directly into the device, based on a single randomly chosen $\hat{U}$. Although this device provides a much higher refractive index contrast than SiO_xN_y, the lumped fibre-to-fibre coupling efficiency was typically much lower—on the order of ∼5 %. We attribute much of this loss to poor mode-matching between fibre and waveguide, rather than propagation loss, and expect that this can be considerably improved.

6.3.3.4 Pseudo-number-Resolving Detection

BosonSampling only requires that measurements are performed in the *collision-free subspace* where no two photons occupy the same mode. It is therefore sufficient to use non-number resolving detectors, such as the silicon APDs used throughout this thesis. For quantum walks, however, the most interesting features occur when photons bunch together i.e. on the main diagonal of the correlation matrix ($i \approx j \approx k \ldots$). In order to observe these effects, we must be able to count up to four photons in single mode. While number-resolving detectors have recently been reported both at room temperature and using superconducting nanowires (see Sect. 1.6.4), they are currently not widely available.

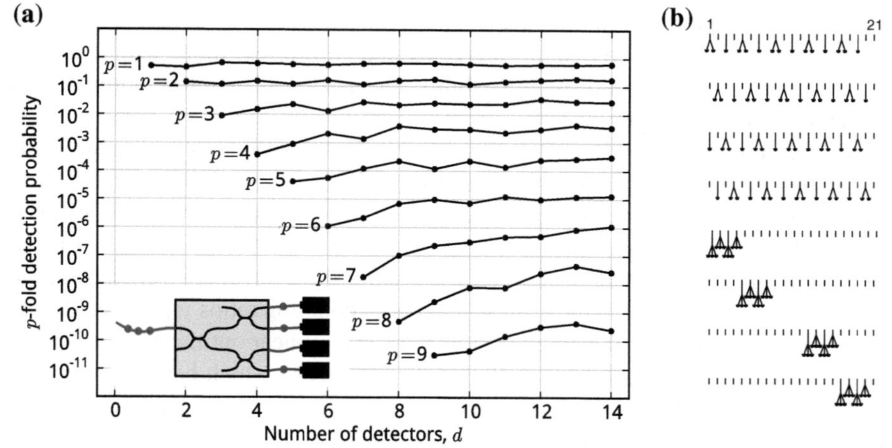

Fig. 6.7 **a** Numerical simulation of pseudo-number-resolving detection efficiency using fibre splitters and multiplexed non-number-resolving detectors. The simulation assumes an average single-detector quantum efficiency of $60 \pm 10\,\%$, and a variance in splitting ratio of $\sim 10\,\%$—realistic experimental values. Inset: an example with $p = 3$, $d = 4$. **b** Eight detection schemes used to image three-photon data in a quantum walk

In order to examine the collision subspace of probability distributions generated by the quantum walk chip, we instead multiplex silicon APDs using fibre splitters, thus approximating non-deterministic number-resolving detectors. Using d unit-efficiency detectors, together with a balanced 1-to-d fibre splitter, we ideally detect p photons in a single mode with probability

$$P(p, d) = \binom{d}{p} / \left[\binom{d+p-1}{p} p^p \right]. \tag{6.17}$$

$P(d, p)$ is polynomial in p if $d \geq p^2$, and this scheme is in principle scalable. Numerical simulations of realistic detection efficiencies, taking into account various experimental imperfections, are shown in Fig. 6.7.

6.3.4 *Characterization and Numerical Simulation*

In order to compare our experimental results with theory, we implemented detailed numerical simulations of each setup, incorporating various measured experimental parameters.

The visibility of quantum interference of the photon source was characterized by measurement of Hong-Ou-Mandel dips. Fitting curves to measured count rates as described in Sect. 2.4, we estimated the HOM dip visibility between photons generated in separate downconversion events ("off-pair" photons) to be $\sim 88\,\%$. The visibility of off-pair quantum interference is reduced with respect to an on-pair dip by the possibility that two photons are generated at different times within a single pulse, leading to temporal distinguishability. Our laser was therefore optimized for generation of short pulses.

Fabrication of both QW and RU chips is subject to imperfection in coupling ratios and phase shifts, and the unitary $\hat{U}_d$ describing each device will differ slightly from the $\hat{U}$ originally designed. Owing to the ordered structure of the QW chip, we were able to characterize $\hat{U}_{QW}$ by means of single photon measurements only, using bright laser light injected at the centre of the array. Assuming that deviation from the original design is most prominent in the nearest-neighbour coupling ratios γ_{ij} (which depend exponentially on distance) and time parameter t, a nonlinear optimization algorithm was used to find values of these $(20+1)$ free parameters which best reproduce the experimentally measured single-photon distribution. We found a standard deviation in the reconstructed coupling ratio of $\sigma_\beta \sim 5\,\%$.

For the RU chip, since fabrication error could potentially lead to *any* m-mode unitary, we use the more rigorous approach of Laing [67], which allows full reconstruction of the device unitary by means of a single-photon and two-photon measurements only. This method, which does not require interferometric stability between the chip and probes, is scalable: since $\hat{U}_d$ is described by a number of parameters polynomial in m, it can be completely reconstructed using a polynomial number of measurements.

Our numerical simulations also make use of a full audit of individual detector efficiencies, fibre splitter coupling ratios, and losses, together with a model of each pseudo-number-resolving detection scheme.

6.3.5 Experimental Results

6.3.5.1 Bunching and Clouding in Quantum Walks

In our first experiment, we injected the three-photon state $|111\rangle$ into the central waveguides ($k=10$, 11, 12) of the QW chip, using the fourth output mode of the source as a herald as previously described. Using 1-to-2 and 1-to-3 fibre splitters in a total of eight configurations (Fig. 6.7), we measured 524 of the 1771 possible three-photon detection events over 21 modes, obtaining a total of 3870 threefold events. Delaying the arrival time of photons from modes E and S of the source on the order of the photon coherence time ($\sim$1 ps), we repeated this measurement with mutually distinguishable photons, obtaining 5588 threefold events.

We found a statistical fidelity between normalized theoretical P_i^{th} and experimental P_i^{exp} probability distributions of $F_Q = 0.930 \pm 0.003$ and $F_C = 0.961 \pm 0.002$ for indistinguishable and distinguishable input states respectively. Error bars are calculated using a Monte-Carlo technique, assuming Poissonian statistics. We attribute the observed discrepancy between experiment and theory to imperfect characterization of the QW device, non-uniform facet/coupling loss, limited visibility of quantum interference due to photon distinguishability introduced by propagation through device itself, and higher-order terms in the SPDC state.

Experimental QW data is compared with numerical simulations in Fig. 6.8c, d, g and h. Using indistinguishable photons, bosonic bunching is immediately apparent along the main diagonal of the correlation cube, with three-photon detection events strongly suppressed in off-diagonal regions. Two "clouds" are clearly visible, centred on waveguides 6 and 16. If one photon is detected at waveguide 16 (for example), it is much more likely that the remaining two photons will also be detected in the vicinity of that waveguide. In contrast, using distinguishable photons we are equally likely to detect photons at opposite sides of the array as to find them grouped together, and the clouds are seen to dissipate.

We can compare this behaviour with three-photon data obtained from the unstructured RU chip, shown together with numerical simulations in Fig. 6.8a, b, e and f. For the QW chip, it is meaningful for two waveguides to be nearest neighbours, while for the RU chip this is not the case. No clouding behaviour is observed, and the distinction between distinguishable and indistinguishable photons is qualitatively not as strong.

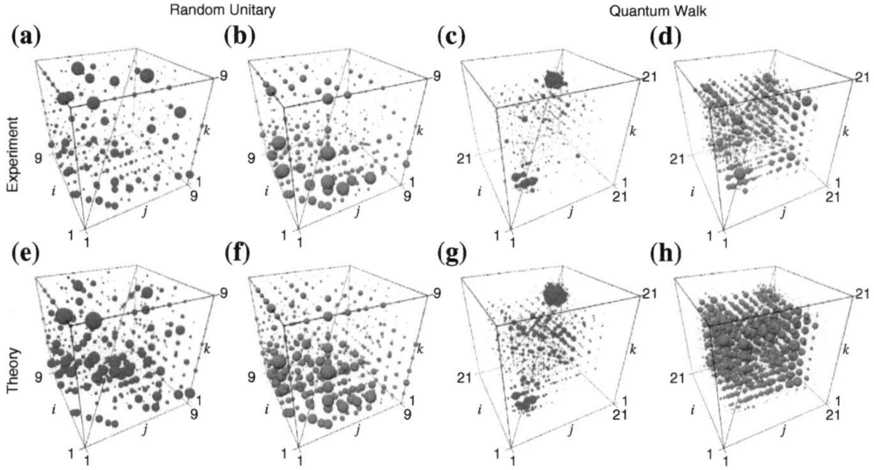

Fig. 6.8 Absence and emergence of correlated bosonic clouds. Three-photon data for a nine mode random unitary (RU, **a**, **b**, **e**, **f**) and a 21 mode quantum walk (QW, **c**, **d**, **g**, **h**). The radii of spheres centred at coordinates (i, j, k) are proportional to the probability of detecting three photons in output modes i, j and k respectively. We tune between indistinguishable (*blue*) and distinguishable (*red*) photons by introducing a large path-length difference at the source. **a** Experimental *RU* with indistinguishable and (**b**) distinguishable photons. (**c**, **d**) Bosonic clouds from experimental *QW* using indistinguishable and distinguishable photons respectively. **e**, **f** Simulated **RU** with indistinguishable and distinguishable photons respectively. **g**, **h** Theoretical bosonic clouds from *QW* with indistinguishable and distinguishable photons respectively. Experimental data has been corrected for measured detection efficiency. Numerics have been filtered to show only those detection patterns which were experimentally measured—this is the main reason for the apparent asymmetry between boson clouds (colour online)

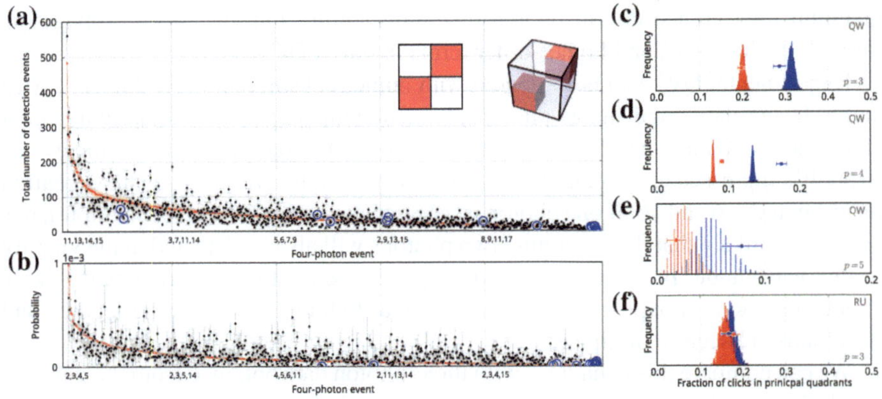

Fig. 6.9 Quantum-walk-specific verification. **a** Experimental data (*black points*) for four indistinguishable photons in a 21 mode quantum walk, over 1820 fourfold detection patterns, ordered by descending theoretical probability (*red points*). Number-resolved data is highlighted with blue circles. Error bars assume Poissonian statistics. **b** Reconstructed pure-state four-photon data, after subtraction of experimentally-measured contributions due to $|2200\rangle$ and $|0022\rangle$ terms. In (**c**–**e**) we perform a quantum-walk-specific test for $p = 3, 4, 5$ photons, measuring the fraction of events C in the principal quadrants (see inset). We plot experimental results for indistinguishable (*blue*) and distinguishable (*red*) photons, along with a corresponding theoretical distribution with the same number of samples drawn. In all cases, we see a statistically significant increase in C for indistinguishable photons. In (**f**) we perform the same test for three photons in a 9-mode random unitary, where our quantum-walk-specific test does not reveal statistically significant quantum-classical separation, as expected

In order to quantify this bosonic clouding effect, we construct a simple metric. For a general experiment of p photons in m modes, the correlation matrix forms a p-dimensional hypercube with 2^p quadrants.[11] We define the *clouding parameter* C to be the fraction of events which occupy the two principal quadrants, i.e. those which intersect the main diagonal $i = j = k = l \ldots$. We obtained experimental values of $C_Q^{\text{exp}} = 0.288 \pm 0.015$ and $C_C^{\text{exp}} = 0.20 \pm 0.01$ for indistinguishable and distinguishable photons respectively, compared to theoretical values of $C_Q^{\text{th}} = 0.332 \pm 0.007$ and $C_C^{\text{th}} = 0.202 \pm 0.005$, indicating significantly stronger clouding under the influence of quantum interference.

Isolating the $|1111\rangle$ term from $|\Psi_{\text{SPDC}}\rangle$ as previously described, we measured four-photon correlations at the output of the QW device, with modes N, S, E and W of the source connected to waveguides ($k = 9, 10, 11, 12$) respectively. Using 1-by-4 splitters in two different configurations, we measured coincidence countrates for 1016 out of a possible 10,626 fourfold patterns, collecting ~50,000 events over the course of ~1 week. Experimental data is plotted together with numerical simulations in Fig. 6.9a, b. We found statistical fidelities between normalized experimental and theoretical distributions of $F_Q = 0.971 \pm 0.001$ and $F_C = 0.978 \pm 0.004$

[11]Higher-dimensional quadrants of hypercubes are referred to as *octants* or *hyper-octants*.

respectively. Experimental and theoretical clouding parameters were measured to be $C_Q^{\text{ex}} = 0.175 \pm 0.007$ and $C_Q^{\text{th}} = 0.144 \pm 0.002$ respectively. In contrast, we measured significantly smaller values of C when all four photons were made distinguishable, finding experimental and theoretical values of $C_C^{ex} = 0.09 \pm 0.003$ and $C_C^{th} = 0.078 \pm 0.001$ respectively. These values are compared graphically in Fig. 6.9d.

While acquiring this four-photon data, the counting system also recorded 217 five-fold detection events. These events arise from extremely low-probability six-photon terms in $|\Psi_{\text{SPDC}}\rangle$, where one photon is lost. The Hilbert space dimension of 5 photons in 21 modes—the number of possible detection patterns—is 53,130. As a result, with so few detection events registered in total, no unique detection pattern appears more than twice in our data. In this regime it is no longer helpful to plot individual countrates in a bar chart, or compute statistical fidelities. However, the clouding metric, which boils the full dataset down to a single global property of the probability distribution, appears to detect evidence of bosonic clouding, and therefore quantum interference, in our experimental data. We measured values of $C_Q = 0.079 \pm 0.019$ and $C_C = 0.058 \pm 0.016$ for indistinguishable and distinguishable photons respectively. These values are compared graphically in Fig. 6.9e.

6.3.5.2 Quantum Verification in Large Hilbert Spaces

Full quantum state tomography (Sect. 2.6) of the three-photon, 21-mode state shown in Fig. 6.8 would require $O(1 \times 10^6)$ measurements to reconstruct the $d^2 - 1$ free parameters of the density matrix $\hat{\rho}$. Without exploiting known structure in the state, full reconstruction of the 5-photon state measured in Fig. 6.9e would require $O(1 \times 10^9)$ linearly independent measurement settings. Even estimating the expectation value of a single measurement setting is likely to be prohibitively time-consuming, as the probability of any given event is so small.

We can compare this situation to that of Shor's algorithm, which is designed in such a way that, for a sufficiently large problem size, the experimentalist *cannot accurately measure the probability of detecting any given n-qubit state* in polynomial time. Specifically, when factoring an L-bit number N, Shor's algorithm generates a periodic probability distribution characterized by $O(N) \propto O(2^L)$ equally spaced peaks. Although it is exponentially more likely that the machine will output a result corresponding to a peak than a trough, since each *peak* is exponentially small, the probability of registering the *same* outcome twice is negligible. Despite this, the *period*—a global property of the probability distribution, which is a function of its highly structured nature—can be extracted (using the inverse quantum Fourier transform QFT) after only polynomially many trials, yielding the prime factors and thus a simple means of verifying the output.

As experiments in quantum computation and quantum information continue to scale in complexity and Hilbert space dimension, the available experimental data will necessarily be increasingly sparse, to the extent that standard methods of comparison with theory will break down. Moreover, we are already approaching the point at which both full numerical simulation of the experimental setup, as well as full characterization by quantum state tomography, become classically intractable. Our results begin to encroach on this regime of extremely sparse data and challenging classical simulation. However, as we have shown, using global measures which exploit known structure in the probability distribution or experimental setup, we are nonetheless able to verify that the machine operates as desired.

This global, structured approach is possible for the QW chip, as the device is specifically designed to generate highly structured probability distributions. How can we confirm successful operation of the RU chip, which is nominally completely unstructured?

6.3.5.3 Experimental Verification of BosonSampling

Injecting three photons into the first three modes of the RU chip, we measured 434 threefold coincidences, distributed over all 84 detection patterns in the collision-free subspace. Experimental results from both indistinguishable and distinguishable photons are compared with numerical simulations (based on the reconstructed experimental device unitary $\hat{U}_d$) in Fig. 6.8a, b, e and f respectively. Statistical fidelities between the experimental data and numerical model were $F_Q = 0.939 \pm 0.010$ and $F_C = 0.970 \pm 0.007$, for indistinguishable and distinguishable photons respectively.

The principal claim of Ref. [62] is that without knowledge of $\hat{U}_d$, the experimentalist cannot discriminate between an untrusted BosonSampling machine and a classical uniform-sampler $\mathcal{F}$, without first measuring an exponential number of samples. In our first approach to verification of the RU device, we assert that in the context of realistic experiments it seems unreasonable to enforce the condition that $\hat{U}_d$ should be unavailable to the experimentalist: as we have already described, it can always be efficiently measured [67]. Indeed, Aaronson and Arkhipov have shown [68] that given $\hat{U}_d$, a BosonSampling machine *can* be distinguished from a uniform-sampling machine in polynomial time using the so-called *row-norm* or R^* discriminator, prompting experimental interest [69].

Sending p photons into modes z_i^a of a device with a known transfer matrix $\Lambda \leftrightarrow \hat{U}_d$, we sample a single detection event, registering a coincidence-click at output modes z_j^b. We isolate the $p \times p$ submatrix M of Λ, choosing columns and rows according to z_i^a and z_j^b respectively, and then compute the normalized product R^* of row-norms of M,

$$R^* = \frac{1}{p^p} \prod_{i=1}^{p} \left(\sum_{j=1}^{p} |M_{ij}|^2 \right). \tag{6.18}$$

Note that this quantity can be computed in classical polynomial time. While R^* does not give a good approximation to $|\operatorname{per}(M)|^2$, it is nonetheless sufficiently *correlated* with $\mathcal{B}$—which *does* depend on $|\operatorname{per}(M)|^2$—to discriminate between BosonSampling devices and uniform-samplers.

In order to confirm that $\mathcal{B}$ as generated by our experiment can rapidly be distinguished from $\mathcal{F}$, we use Bayesian inference to update our knowledge in real time, based on the data shown in Fig. 6.8. Bayes' theorem gives the probability that we are sampling from $\mathcal{B}$, given our experimental values of R^*

$$P(\mathcal{B}|R^*) = \frac{P(R^*|\mathcal{B})P(\mathcal{B})}{P(R^*)}. \tag{6.19}$$

In order to obtain $P(R^*|\mathcal{B})$, we numerically estimate the probability that R^* is above a threshold value of 1, finding $P((R^* > 1)|\mathcal{B}) = 0.631$, $P((R^* < 1)|\mathcal{B}) = 0.369$. Starting from an unbiased prior, $P(\mathcal{B}) = P(\mathcal{F}) = 1/2$, after only 12 detection events we obtain a confidence level greater than 90 % that the experimental data was drawn from $\mathcal{B}$. Using all 434 detection events, this rises to $P(\mathcal{B}|R^*) = 1 - 10^{-35}$.

In $\mathcal{F}$, Goglin et al. consider a somewhat artificial failure mode of a BOSONSAMPLING device—in reality, the experimentalist is likely to know a priori that the device in the lab is not a uniform sampler. A more realistic possibility is that photons sent into the device are partially or completely mutually distinguishable, a legitimate experimental concern. In this case no quantum interference is observed in the output probability distribution $\mathcal{C}$, and the behaviour of the device can be classically predicted in polynomial time. Here, the R^* test fails to distinguish $\mathcal{B}$ from a classical machine generating $\mathcal{C}$. As we have already seen, the clouding metric C also fails to detect a signature of quantum interference in BOSONSAMPLING data, owing to the lack of structure in $\mathcal{B}$.

An alternative test, which succeeds in this task, measures the *net probability* that a p-fold click is detected in the collision-free subspace, when p photons are sent into the device. Intuitively, since indistinguishable photons tend to bunch together, this probability should increase when input photons are made distinguishable. The fraction of trials N to p-fold detection events P was estimated to be $P_Q^{ex}(p-\text{fold}) = 0.450 \pm 0.028$ and $P_C^{ex}(p-\text{fold}) = 0.680 \pm 0.002$ for indistinguishable and distinguishable photons respectively, compared to theoretical values of $P_Q^{th} = 0.509$ and $P_C^{th} = 0.691$. Here we used the method of [69] to estimate N.

6.3.6 Postselected Multiphoton Quantum Walks

One of the most powerful features of BOSONSAMPLING is that it demands neither postselection nor adaptive measurement. It is remarkable that BOSONSAMPLING provides a quantum speedup in the absence of any nonlinear coupling between photons, either in the sense of nonlinear optics (Sect. 1.5.5) or the measurement-induced nonlinearity of KLM (Sect. 1.6.2). It is nonetheless interesting to ask whether anything might be accomplished by minimal postselection and/or feed-forward on quantum walks or BOSONSAMPLING machines.

The probability distribution generated by a quantum walk of a single photon in a linear, uniformly coupled array is symmetric about the input waveguide (Fig. 6.4). For multiphoton walks, if the choice of input waveguides is symmetric, the multi-photon distribution will also be symmetric, as seen in our three-photon experimental data (Fig. 6.8). However, by *postselecting* on detection of one photon in a particular waveguide, we find that interesting asymmetric effects can be seen in the resulting $(p-1)$-photon statistics.

Figure 6.10a–f show numerical simulations of the time evolution of a four-photon quantum walk, after postselection on detection of one photon in a particular off-centre mode. The resulting three-photon statistics show an asymmetric distribution, with a single ballistic lobe propagating on the main diagonal. We expect that this

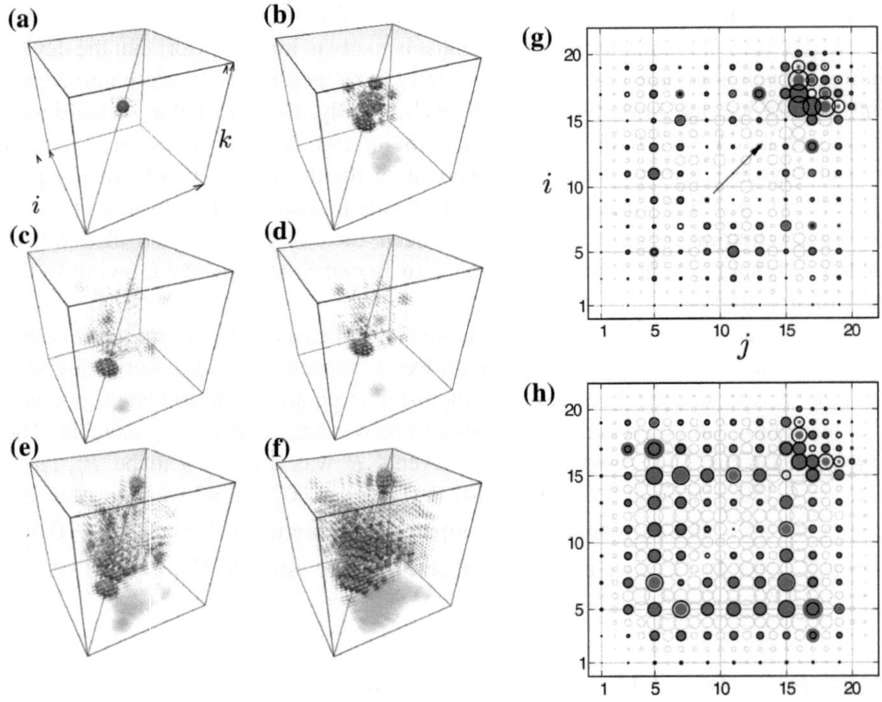

Fig. 6.10 Asymmetry in postselected on quantum walks. **a–f** Numerical simulation of quantum walk time evolution under postselection. Beginning with a four-photon quantum walk, we postselect on detection of one photon in a specific waveguide. Figures (**a–f**) show successive steps in the time evolution of the resulting three photon state as correlation cubes. Each axis of the cube denotes the position of a photon in the array, where the hue and radius of each sphere are proportional to the probability of detecting three photons, after postselection, at waveguides i, j, k. **f** A two-photon quantum walk, postselected from three-photon data. The radius of each red circle corresponds to the experimental count rate in waveguides i, j, after postselection and correction for measured detection efficiencies. Black circles show a numerical simulation. The asymmetry seen in the numerics (**a–f**) is clearly reproduced. **h** Experimental data using distinguishable photons. The apparent asymmetry is an artefact of our measurement setup

effect would be difficult to achieve without postselection, assuming a uniform, planar waveguide array.[12] We used the QW chip to test this behaviour, sending three indistinguishable photons into the device as before and postselecting on detection at waveguide 15. Experimental two-photon correlations are shown in Fig. 6.10g, where a single asymmetric lobe can be clearly seen. Using distinguishable photons, we do not see the same effect (Fig. 6.10h).

6.3.7 Discussion

The experimental progress described in this section is characterized by an increase in complexity. Each optical chip has more than twice as many spatial modes as those previously described in this thesis, and the RU device has 36 directional couplers, compared to 13 for the CNOT-MZ. We have described a four-photon source which, although not entirely novel, has been used in a previously unexplored capacity. To our knowledge, ours is the first demonstration of correlated coincidence counting using 16 single photon detectors where all possible detection events are registered.[13] This system has allowed us to take detailed images of the complex three-photon interference effects shown in Fig. 6.8, which have not previously been observed.

In [68], Aaronson and Arkhipov describe an efficient classical algorithm due to Fernando Brandao, based on work of Trevisan et al. [70], which generates a "mock-up" probability distribution $\mathcal{M}$ which provably cannot be distinguished from $\mathcal{B}$ by circuits of any fixed polynomial size. This distribution is carefully designed , and it is not currently known whether any machine—classical or quantum—could distinguish $\mathcal{M}$ from $\mathcal{B}$ in polynomial time. Although it is hard to say at this stage, it would not be surprising if the *deliberate* lack of structure in $\mathcal{B}$ renders BosonSampling machines fundamentally indistinguishable from certain adversarial classical "fakes". Nonetheless, we have shown that a specific small class of experimentally relevant BosonSampling failure modes can be efficiently detected in experiments. We expect that the scope of such methods will grow, to encompass the majority of realistic errors that might render BosonSampling machines classically tractable.

Moreover, we have found circumstantial evidence to suggest that by deliberately imposing *structure* on the interferometer and resulting probability distribution, the difficulty of experimental verification can be significantly reduced. To this end, we have shown that by exploiting known qualitative properties of the probability distribution generated by a quantum walk, we are able to detect a signature of quantum interference even in extremely large and non-separable Hilbert spaces, where it is no longer practical to measure probabilities. We do not expect that the scheme used in Sect. 6.3.5, as described, will always return a definitive answer after a polynomial

---

[12]Careful control of the phase of input photons, following the classical approach of beam steering using a phased array, might conceivably reproduce this effect.

[13]16 detectors are used in Ref. [60], but only a subset of possible detection events are recorded.

number of events.[14] However, we expect that future scalable techniques will follow our basic method, and that probability distributions with tailored structure will be essential to light the way, as we drive out into the darkness of classical computational intractability.

6.4 Statement of Work

The majority of my contribution to work described in this section has been in the construction and programming of the counting system (Sect. 6.2), as well as analysis of multiphoton data. I also performed many of the numerical simulations, and worked on data acquisition in the lab.

References

1. J.C.F. Matthews, R. Whittaker, J.L. O'Brien, P.S. Turner, Testing randomness with photons (2013). arXiv:1312.1940
2. A. Muthukrishnan, C.R. Stroud Jr, Multivalued logic gates for quantum computation. Phys. Rev. A **62**(5), 052309 (2000)
3. X.-Q. Zhou, T.C. Ralph, P. Kalasuwan, M. Zhang, A. Peruzzo, B.P. Lanyon, J.L. O'Brien, Adding control to arbitrary quantum operations (2010)
4. J. Sun, E. Timurdogan, A. Yaacobi, E.S. Hosseini, M.R. Watts, Large-scale nanophotonic phased array. Nature **493**, 195 (2013)
5. S. Aaronson, A. Arkhipov, The computational complexity of linear optics (2010)
6. G.M. Viswanathan, S.V. Buldyrev, S. Havlin, M.G.E. da Luz, E.P. Raposo, H.E. Stanley, Optimizing the success of random searches. Nature **401**, 911–914 (1999)
7. D.A. Raichlen, B.M. Wood, A.D. Gordon, A.Z.P. Mabulla, F.W. Marlowe, H. Pontzer, Evidence of lévy walk foraging patterns in human hunter-gatherers. Proc. Nat. Acad. Sci. (2013)
8. R. Motwani, P. Raghvan, *Randomized Algorithms* (Cambridge University Press, Cambridge, 1995)
9. H.W. Li, J. Wabnig, D. Bitauld, P. Shadbolt, A. Politi, A. Laing, J.L. O'Brien, A.O. Niskanen, Calibration and high fidelity measurement of a quantum photonic chip. New J. Phys. **15**(6), 063017 (2013)
10. M. Jerrum, A. Sinclair, E. Vigoda, A polynomial-time approximation algorithm for the permanent of a matrix with nonnegative entries. J. ACM **51**(4), 671–697 (2004)
11. M. Mohseni, P. Rebentrost, S. Lloyd, A. Aspuru-Guzik, Environment-assisted quantum walks in photosynthetic energy transfer. J. Chem. Phys. **129**(17), 174106 (2008)
12. J. Klafter, R. Silbey, Phys. Lett. **76A**, 143 (1980)
13. A.M. Childs, J. Goldstone, Spatial search by quantum walk. Phys. Rev. A **70**(2), 022314 (2004)
14. N. Shenvi, J. Kempe, K.B. Whaley, Quantum random-walk search algorithm. Phys. Rev. A **67**(5), 052307 (2003)
15. A.M. Childs, J.M. Eisenberg, Quantum algorithms for subset finding (2003). arXiv:quant-ph/0311038
16. H. Buhrman, R. Spalek, Quantum verification of matrix products (2004). arXiv:quant-ph/0409035

[14]Doubts might be raised by the fact that the number of hypercube orthants which intersect with the main diagonal of the correlation matrix falls off exponentially with p.

17. E. Farhi, J. Goldstone, S. Gutmann, A quantum algorithm for the Hamiltonian NAND Tree (2007). arXiv:quant-ph/0702144
18. B.W. Reichardt, R. Spalek. Span-program-based quantum algorithm for evaluating formulas (2007). arXiv:0710.2630
19. A. Childs, Universal computation by quantum walk. Phys. Rev. Lett. **102** (2009)
20. A.M. Childs, D. Gosset, Z. Webb, Universal computation by multiparticle quantum walk. Science **339**, 791–794 (2013)
21. R.P. Feynman, Simulating physics with computers. Int. J. Theor. Phy. Theor. Phy. **21**, 467–488 (1982)
22. D. Aharonov, A. Ambainis, J. Kempe, U. Vazirani, Quantum walks on graphs (2000). arXiv:quant-ph/0012090
23. A. Nayak, A. Vishwanath, Quantum walk on the line (2000). arXiv:quant-ph/0010117
24. J. Watrous, Quantum simulations of classical random walks and undirected graph connectivity (1998). arXiv:cs/9812012
25. A.M. Childs, E. Farhi, S. Gutmann, An example of the difference between quantum and classical random walks (2001). arXiv:quant-ph/0103020
26. E. Farhi, S. Gutmann, Quantum computation and decision trees. Phys. Rev. A **58**, 915–928 (1998)
27. A. Childs, On the relationship between continuous-and discrete-time quantum walk. Commun. Math. Phys. (2010)
28. M.C. Rechtsman, J.M. Zeuner, A. Tünnermann, S. Nolte, M. Segev, A. Szameit, Strain-induced pseudomagnetic field and photonic Landau levels in dielectric structures. Nat. Photonics **7**, 153–158 (2013)
29. J.D.A. Meinecke, K. Poulios, A. Politi, J.C.F. Matthews, A. Peruzzo, N. Ismail, K. Wörhoff, J.L. O'Brien, M.G. Thompson, Coherent time evolution and boundary conditions of two-photon quantum walks in waveguide arrays. Phys. Rev. A **88**(1), 012308 (2013)
30. M. Karski, L. Förster, J.-M. Choi, A. Steffen, W. Alt, D. Meschede, A. Widera, Quantum walk in position space with single optically trapped atoms. Science **325**(5937), 174–177 (2009)
31. H. Schmitz, R. Matjeschk, C. Schneider, J. Glueckert, M. Enderlein, T. Huber, T. Schaetz, Quantum walk of a trapped ion in phase space. Phys. Rev. Lett. **103**(9), 090504 (2009)
32. F. Zähringer, G. Kirchmair, R. Gerritsma, E. Solano, R. Blatt, C.F. Roos, Realization of a quantum walk with one and two trapped ions. Phys. Rev. Lett. **104**(10), 100503 (2010)
33. J. Du, H. Li, X. Xu, M. Shi, J. Wu, X. Zhou, R. Han, Experimental implementation of the quantum random-walk algorithm. Phys. Rev. A **67**(4), 042316 (2003)
34. D. Bouwmeester, I. Marzoli, G.P. Karman, W. Schleich, J.P. Woerdman, Optical galton board. Phys. Rev. A **61**, 013410 (1999)
35. A. Schreiber, K.N. Cassemiro, V. Potoček, A. Gábris, P.J. Mosley, E. Andersson, I. Jex, Ch. Silberhorn, Photons walking the line: a quantum walk with adjustable coin operations. Phys. Rev. Lett. **104**, 050502 (2010)
36. A. Schreiber, K.N. Cassemiro, V. Potoček, A. Gábris, I. Jex, Ch. Silberhorn, Decoherence and disorder in quantum walks: from ballistic spread to localization. Phys. Rev. Lett. **106**, 180403 (2011)
37. M.A. Broome, A. Fedrizzi, B.P. Lanyon, I. Kassal, A. Aspuru-Guzik, A.G. White, Discrete single-photon quantum walks with tunable decoherence. Phys. Rev. Lett. **104**, 153602 (2010)
38. P. Knight, E. Roldan, Quantum walk on the line as an interference phenomenon. Phys. Rev. A (2003)
39. A. Childs, R. Cleve, E. Deotto, E. Farhi, Exponential algorithmic speedup by a quantum walk. in *Proceedings of STOC'2003* (2003)
40. Y. Omar, N. Paunković, L. Sheridan, Quantum walk on a line with two entangled particles. Phys. Rev. A **74** (2006)
41. A. Peruzzo, M. Lobino, J.C.F. Matthews, N. Matsuda, A. Politi, K. Poulios, X.-Q. Zhou, Y. Lahini, N. Ismail, K. Wörhoff, Y. Bromberg, Y. Silberberg, M.G. Thompson, J.L. OBrien. Quantum walks of correlated photons. Science **329**, 1500–1503 (2010)

42. J.C.F. Matthews, K. Poulios, J.D.A. Meinecke, A. Politi, A. Peruzzo, N. Ismail, K. Wörhoff, M.G. Thompson, J.L. O'Brien. Observing fermionic statistics with photons in arbitrary processes: scientific reports: nature publishing group. Sci. Rep. **3** (2013)
43. L. Sansoni, F. Sciarrino, G. Vallone, P. Mataloni, A. Crespi, R. Ramponi, R. Osellame, Two-particle bosonic-fermionic quantum walk via integrated photonics. Phys. Rev. Lett. **108**, 010502 (2012)
44. A. Crespi, R. Osellame, R. Ramponi, D.J. Brod, E.F. Galvão, N. Spagnolo, C. Vitelli, E. Maiorino, P. Mataloni, F. Sciarrino, Integrated multimode interferometers with arbitrary designs for photonic boson sampling. Nat. Photonics (2013)
45. A. Crespi, R. Osellame, R. Ramponi, V. Giovannetti, R. Fazio, L. Sansoni, F. de Nicola, F. Sciarrino, P. Mataloni, Anderson localization of entangled photons in an integrated quantum walk. Nat. Photonics **7**, 322–328 (2013)
46. K. Poulios, R. Keil, D. Fry, J.D.A. Meinecke, J.C.F. Matthews, A. Politi, M. Lobino, M. Gräfe, M. Heinrich, S. Nolte, A. Szameit, J.L. O'Brien, Quantum walks of correlated photon pairs in two-dimensional waveguide arrays (2013)
47. T. Fukuhara, P. Schauß, M. Endres, S. Hild, M. Cheneau, I. Bloch, C. Gross, Microscopic observation of magnon bound states and their dynamics (2013). arXiv:1305.6598
48. A.J. Bessen, Distributions of continuous-time quantum walks (2006). arXiv:quant-ph/0609128
49. Ian Parberry
50. N. Dershowitz, Y. Gurevich, A natural axiomatization of computability and proof of church's thesis. Bull. Symbolic Logic **14**, (2008)
51. N. Dershowitz, E. Falkovich, A formalization and proof of the extended church-turing thesis. EPTCS (2011)
52. G. Kalai, How quantum computers fail: quantum codes, correlations in physical systems, and noise accumulation (2011)
53. E. Martín-López, A. Laing, T. Lawson, R. Alvarez, X.-Q. Zhou, J.L. O'Brien, Experimental realization of Shor's quantum factoring algorithm using qubit recycling. Nat. Photonics **6**, 773–776 (2012)
54. Xu Nanyang, Jing Zhu, Lu Dawei, Xianyi Zhou, Xinhua Peng, Du Jiangfeng, Quantum factorization of 143 on a dipolar-coupling nuclear magnetic resonance system. Phys. Rev. Lett. **108**, 130501 (2012)
55. S.J. Devitt, K. Nemoto, Programming a topological quantum computer (2012)
56. Alan Aspuru-Guzik, Philip Walther, Photonic quantum simulators. Nat. Phys. **8**(4), 285–291 (2012)
57. L.G. Valiant, The complexity of computing the permanent. Theoret. Comput. Sci. **8**, 189–201 (1979)
58. L. Gurvits, On the complexity of mixed discriminants and related problems. Math. Found. Comput. Sci. 447–458 (2005)
59. H.J. Ryser, Combinatorial mathematics. Carus Math. Monogr. **14** (1963)
60. X.-C. Yao, T.-X. Wang, P. Xu, H. Lu, G.-S. Pan, X.-H. Bao, C.-Z. Peng, C.-Y. Lu , Y.-A. Chen, J.-W. Pan, Observation of eight-photon entanglement. Nat. Photonics **6**(4), 225–228 (2012)
61. A. Leverrier, R. García-Patrón. Does Boson sampling need fault-tolerance? (2013)
62. C. Gogolin, M. Kliesch, L. Aolita, J. Eisert, Boson-sampling in the light of sample complexity (2013)
63. M.A. Broome, A. Fedrizzi, S. Rahimi-Keshari, J. Dove, S. Aaronson, T.C. Ralph, A.G. White, Photonic Boson sampling in a tunable circuit. Science **339**, 794–798 (2013)
64. J.B. Spring, B.J. Metcalf, P.C. Humphreys, W.S. Kolthammer, X.-M. Jin, M. Barbieri, A. Datta, N. Thomas-Peter, N.K. Langford, D. Kundys, J.C. Gates, B.J. Smith, P.G.R. Smith, I.A. Walmsley, Boson sampling on a photonic chip. Science **339**, 798–801 (2013)
65. A. Crespi, R. Osellame, R. Ramponi, D.J. Brod, E.F. Galvao, N. Spagnolo, C. Vitelli, E. Maiorino, P. Mataloni, F. Sciarrino, Experimental boson sampling in arbitrary integrated photonic circuits. Nat. Photonics **7**(7), 545 (2013)
66. M. Tillmann, B. Dakić, R. Heilmann, S. Nolte, A. Szameit, P. Walther, Experimental Boson sampling. Nat. Photonics **7**(7), 540–544 (2013)

67. A. Laing, J.L. O'Brien, Super-stable tomography of any linear optical device. (2012)
68. S. Aaronson, A, Arkhipov, BosonSampling is far from uniform. (2013)
69. N. Spagnolo, C. Vitelli, M. Bentivegna, D.J. Brod, A. Crespi, F. Flamini, S. Giacomini, G. Milani, R. Ramponi, P. Mataloni, R. Osellame, E.F. Galvao, F. Sciarrino, Efficient experimental validation of photonic boson sampling against the uniform distribution. (2013)
70. L. Trevisan, M. Tulsiani, S. Vadhan, Regularity, boosting, and efficiently simulating every high-entropy distribution. in *Proceedings of IEEE Conference on Computational Complexity* (2009)

Chapter 7
Discussion

Finally, I was let down and joined the others at the window, to watch the sleet fall.

Ivor Cutler

We have described a broad spectrum of experiments in quantum photonics, many of which make use of the control and complexity afforded by monolithic integration. In our work with the CNOT-MZ, we have shown the value of reconfigurability in such devices, and the surprising diversity of experiments which can be performed with just two qubits. In doing so, we have confirmed that integrated quantum photonics can reproduce the performance and flexibility of bulk optics.

Using this device we have implemented a new variant on Wheeler's delayed choice experiment, observing continuous tuning between wave and particle phenomena for the first time. While we do not contend that this result provides new physical understanding over and above Bell's theorem, for example, we suggest that it nonetheless provides a useful pedagogical tool to think about wave-particle duality.

In Chap. 4, we introduced three new protocols, which allow the presence of entanglement to be certified under suboptimal experimental conditions. It is reasonable to think that these techniques will be useful for the characterization of quantum states in the laboratory, where calibration and alignment can sometimes be problematic. We believe that these methods might also find applications in quantum key distribution and related quantum communication protocols, when two distant parties do not share a common frame.

Chapter 5 introduced a new algorithm for quantum chemistry. Although the analysis is not complete, we believe that this technique potentially offers very significant benefits over the current status quo for quantum simulation, particularly with respect to the number of gate operations required. Even if this algorithm is not used in the exact form described here, we anticipate that realistic implementations of quantum simulators will need to adopt the pragmatic approach described here. We ran our algorithm using the CNOT-MZ, demonstrating both the ability of the algorithm to simulate larger systems with fewer resources, as well as further testing the performance and repeatability of the integrated quantum chip.

P. Shadbolt, *Complexity and Control in Quantum Photonics*,
Springer Theses, DOI 10.1007/978-3-319-21518-1_7

Finally, Chap. 6 describes a number of technical advances in both state preparation and measurement. As with the CNOT-MZ, we again see that by increasing the number of experimental degrees of freedom by a relatively small amount, we expand the diversity and power of experimental quantum phenomena very significantly. We expect that successful verification techniques for BosonSampling-like problems, if not following exactly the method outlined in Chap. 6, will at least depend on the fundamental ideas described therein: namely, deliberate introduction and exploitation of structure in the device and resulting probability distribution.

Appendix A
QY

In the course of the experimental work described in this thesis we have developed a broad general-purpose base of computer code (qy), which is maintained and documented as a library to encourage re-use. The majority of this code is written in the `Python` programming language,[1] with some compiled extensions written in `C` or `Cython` for speed. Both of these languages are free and open source.

qy includes modules for data acquisition (DAQ) and hardware control, data logging and analysis, and numerical simulation, with a specific emphasis on tasks which often occur in experimental quantum photonics. The code is currently open source, and can be obtained via `git`:

`https://github.com/peteshadbolt/qy`

The top-level structure of the library is as follows:

- `qy.analysis`: Various standard metrics and tools for data analysis.
- `qy.formats`: File formats, in particular an efficient format to represent multi-photon coincidence-counting data.
- `qy.graphics`: Utility functions for graphics and plotting.
- `qy.hardware`: Interfaces to various pieces of standard laboratory apparatus, including FPGA counting systems, the DPC-230 described in Sect. 6.2, Toptica diode lasers, Coherent Ti:Saph lasers, custom powermeters, Thor labs SMC100 power meters, silica-on-silicon thermal phase shifters, etc.
- `qy.settings`: Utility functions to read, write and persist global settings.
- `qy.simulation`: Provides general quantum information primitives, including single-qubit states and operators, frequently used two-qubit states and gate operations, measures such as quantum state fidelity and concurrence, a circuit-model simulator, and an optimized linear-optics simulator, capable of calculating multi-photon states and statistics in arbitrary linear optical networks.

[1] `http://www.python.org/`.

P. Shadbolt, *Complexity and Control in Quantum Photonics*,
Springer Theses, DOI 10.1007/978-3-319-21518-1

- `qy.util`: Utility functions.
- `qy.wx`: Extends the functionality of the `wx` GUI library.

Here we will discuss two components in particular: the `linear_optics` simulation package and the `.counted` file format.

A.1 Universal Linear Optics Simulator

The module `qy.simulation.linear_optics` provides a simple means to simulate multiphoton states and statistics in arbitrary linear optical circuits. This work draws upon ideas and code kindly provided by Jasmin Meinecke, Nick Russell, Jacques Carolan. The numerical method is exactly that described in Sect. 1.5.3, and as such depends almost entirely on the calculation of permanents. We have developed optimized code to compute the permanent of complex matrices, using a number of different algorithms and implementations. We implemented the core algorithm using `Cython`, a compiled language which can typically achieve much better performance than standard `Python`, which is interpreted rather than compiled. Typical real-world performance of these methods is summarized in Fig. A.1. The library is very easy to use:

```
import numpy as np
from qy.simulation import linear_optics as lo

# Load up a device from a JSON definition file:
device=lo.beamsplitter_network(json='devices/cnot_mz.json')
print device
print device.get_unitary().round(2)
print device.nmodes

# Draw the waveguide structure as a PDF file
device.draw('devices/cnot_mz.pdf')

# Make a simulator, and link it to the device
simulator=lo.simulator(device, nphotons=2)

# Print out the basis
print simulator.basis

# Set the input state to two photons in the top mode, and look at
# the output probabilities and output state
simulator.set_input_state([0, 0])
print simulator.input_state
print simulator.get_probabilities().round(2)
print simulator.get_output_state()

# Superposition input states, and classical statistics
state=simulator.basis.get_state()
state[0,1]=1/np.sqrt(2)
state[3,4]=1/np.sqrt(2)
print state
simulator.set_input_state(state)
simulator.set_visibility(0.5)
print simulator.get_probabilities()
```

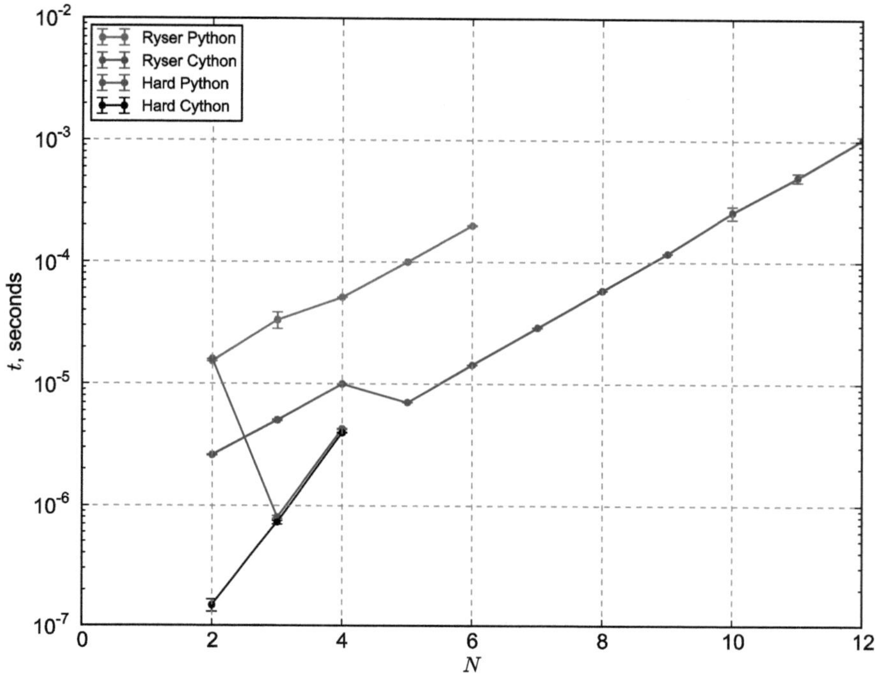

Fig. A.1 Estimated performance of various implementations of algorithms to compute the permanent, tested against 1000 Haar-random $N \times N$ matrices. *Red* and *blue lines* show average execution times for Ryser's algorithm, implemented in Python and Cython respectively, as a function of N. *Green* and *black lines* correspond to execution times for hard-coded implementations *up* to $N = 4$, again in Python and Cython respectively (color online)

```
# Performance test: 4 photons in 16 modes of a Haar-random U
# Hilbert space dimension is now 3876
device=lo.random_unitary(16)
simulator=lo.simulator(device, nphotons=4)
simulator.set_input_state(range(4))  # Photons go in the top 4 modes
probs=simulator.get_probabilities(label=True)
```

When computing the permanent, it was noticed (by Nick Russell) that hard-coded routines can give a significant advantage in speed for small matrices, as the overhead associated with loops and conditional statements can be completely avoided. For completeness we include code up to $N = 4$, beyond which the advantage with respect to Ryser's algorithm is negligible.

```
def perm_2x2(a):
    """ An explicit 2x2 permanent """
    return a[0,0]*a[1,1]
        + a[1,0]*a[0,1]

def perm_3x3(a):
    """ An explicit 3x3 permanent """
    return a[0,0]*a[1,1]*a[2,2]
        + a[0,0]*a[2,1]*a[1,2]
        + a[1,0]*a[0,1]*a[2,2]
        + a[1,0]*a[2,1]*a[0,2]
        + a[2,0]*a[0,1]*a[1,2]
        + a[2,0]*a[1,1]*a[0,2]
```

```
def perm_4x4(a):
    """ An explicit 4x4 permanent """
    return a[0,0]*a[1,1]*a[2,2]*a[3,3]  + a[0,0]*a[1,1]*a[3,2]*a[2,3]
         + a[0,0]*a[2,1]*a[1,2]*a[3,3]  + a[0,0]*a[2,1]*a[3,2]*a[1,3]
         + a[0,0]*a[3,1]*a[1,2]*a[2,3]  + a[0,0]*a[3,1]*a[2,2]*a[1,3]
         + a[1,0]*a[0,1]*a[2,2]*a[3,3]  + a[1,0]*a[0,1]*a[3,2]*a[2,3]
         + a[1,0]*a[2,1]*a[0,2]*a[3,3]  + a[1,0]*a[2,1]*a[3,2]*a[0,3]
         + a[1,0]*a[3,1]*a[0,2]*a[2,3]  + a[1,0]*a[3,1]*a[2,2]*a[0,3]
         + a[2,0]*a[0,1]*a[1,2]*a[3,3]  + a[2,0]*a[0,1]*a[3,2]*a[1,3]
         + a[2,0]*a[1,1]*a[0,2]*a[3,3]  + a[2,0]*a[1,1]*a[3,2]*a[0,3]
         + a[2,0]*a[3,1]*a[0,2]*a[1,3]  + a[2,0]*a[3,1]*a[1,2]*a[0,3]
         + a[3,0]*a[0,1]*a[1,2]*a[2,3]  + a[3,0]*a[0,1]*a[2,2]*a[1,3]
         + a[3,0]*a[1,1]*a[0,2]*a[2,3]  + a[3,0]*a[1,1]*a[2,2]*a[0,3]
         + a[3,0]*a[2,1]*a[0,2]*a[1,3]  + a[3,0]*a[2,1]*a[1,2]*a[0,3]

def perm_5x5(a):
    """ An explicit 5x5 permanent """
    return a[0,0]*a[1,1]*a[2,2]*a[3,3]*a[4,4]  + a[0,0]*a[1,1]*a[2,2]*a[4,3]*a[3,4]
         + a[0,0]*a[1,1]*a[3,2]*a[2,3]*a[4,4]  + a[0,0]*a[1,1]*a[3,2]*a[4,3]*a[2,4]
         + a[0,0]*a[1,1]*a[4,2]*a[2,3]*a[3,4]  + a[0,0]*a[1,1]*a[4,2]*a[3,3]*a[2,4]
         + a[0,0]*a[2,1]*a[1,2]*a[3,3]*a[4,4]  + a[0,0]*a[2,1]*a[1,2]*a[4,3]*a[3,4]
         + a[0,0]*a[2,1]*a[3,2]*a[1,3]*a[4,4]  + a[0,0]*a[2,1]*a[3,2]*a[4,3]*a[1,4]
         + a[0,0]*a[2,1]*a[4,2]*a[1,3]*a[3,4]  + a[0,0]*a[2,1]*a[4,2]*a[3,3]*a[1,4]
         + a[0,0]*a[3,1]*a[1,2]*a[2,3]*a[4,4]  + a[0,0]*a[3,1]*a[1,2]*a[4,3]*a[2,4]
         + a[0,0]*a[3,1]*a[2,2]*a[1,3]*a[4,4]  + a[0,0]*a[3,1]*a[2,2]*a[4,3]*a[1,4]
         + a[0,0]*a[3,1]*a[4,2]*a[1,3]*a[2,4]  + a[0,0]*a[3,1]*a[4,2]*a[2,3]*a[1,4]
         + a[0,0]*a[4,1]*a[1,2]*a[2,3]*a[3,4]  + a[0,0]*a[4,1]*a[1,2]*a[3,3]*a[2,4]
         + a[0,0]*a[4,1]*a[2,2]*a[1,3]*a[3,4]  + a[0,0]*a[4,1]*a[2,2]*a[3,3]*a[1,4]
         + a[0,0]*a[4,1]*a[3,2]*a[1,3]*a[2,4]  + a[0,0]*a[4,1]*a[3,2]*a[2,3]*a[1,4]
         + a[1,0]*a[0,1]*a[2,2]*a[3,3]*a[4,4]  + a[1,0]*a[0,1]*a[2,2]*a[4,3]*a[3,4]
         + a[1,0]*a[0,1]*a[3,2]*a[2,3]*a[4,4]  + a[1,0]*a[0,1]*a[3,2]*a[4,3]*a[2,4]
         + a[1,0]*a[0,1]*a[4,2]*a[2,3]*a[3,4]  + a[1,0]*a[0,1]*a[4,2]*a[3,3]*a[2,4]
         + a[1,0]*a[2,1]*a[0,2]*a[3,3]*a[4,4]  + a[1,0]*a[2,1]*a[0,2]*a[4,3]*a[3,4]
         + a[1,0]*a[2,1]*a[3,2]*a[0,3]*a[4,4]  + a[1,0]*a[2,1]*a[3,2]*a[4,3]*a[0,4]
         + a[1,0]*a[2,1]*a[4,2]*a[0,3]*a[3,4]  + a[1,0]*a[2,1]*a[4,2]*a[3,3]*a[0,4]
         + a[1,0]*a[3,1]*a[0,2]*a[2,3]*a[4,4]  + a[1,0]*a[3,1]*a[0,2]*a[4,3]*a[2,4]
         + a[1,0]*a[3,1]*a[2,2]*a[0,3]*a[4,4]  + a[1,0]*a[3,1]*a[2,2]*a[4,3]*a[0,4]
         + a[1,0]*a[3,1]*a[4,2]*a[0,3]*a[2,4]  + a[1,0]*a[3,1]*a[4,2]*a[2,3]*a[0,4]
         + a[1,0]*a[4,1]*a[0,2]*a[2,3]*a[3,4]  + a[1,0]*a[4,1]*a[0,2]*a[3,3]*a[2,4]
         + a[1,0]*a[4,1]*a[2,2]*a[0,3]*a[3,4]  + a[1,0]*a[4,1]*a[2,2]*a[3,3]*a[0,4]
         + a[1,0]*a[4,1]*a[3,2]*a[0,3]*a[2,4]  + a[1,0]*a[4,1]*a[3,2]*a[2,3]*a[0,4]
         + a[2,0]*a[0,1]*a[1,2]*a[3,3]*a[4,4]  + a[2,0]*a[0,1]*a[1,2]*a[4,3]*a[3,4]
         + a[2,0]*a[0,1]*a[3,2]*a[1,3]*a[4,4]  + a[2,0]*a[0,1]*a[3,2]*a[4,3]*a[1,4]
         + a[2,0]*a[0,1]*a[4,2]*a[1,3]*a[3,4]  + a[2,0]*a[0,1]*a[4,2]*a[3,3]*a[1,4]
         + a[2,0]*a[1,1]*a[0,2]*a[3,3]*a[4,4]  + a[2,0]*a[1,1]*a[0,2]*a[4,3]*a[3,4]
         + a[2,0]*a[1,1]*a[3,2]*a[0,3]*a[4,4]  + a[2,0]*a[1,1]*a[3,2]*a[4,3]*a[0,4]
         + a[2,0]*a[1,1]*a[4,2]*a[0,3]*a[3,4]  + a[2,0]*a[1,1]*a[4,2]*a[3,3]*a[0,4]
         + a[2,0]*a[3,1]*a[0,2]*a[1,3]*a[4,4]  + a[2,0]*a[3,1]*a[0,2]*a[4,3]*a[1,4]
         + a[2,0]*a[3,1]*a[1,2]*a[0,3]*a[4,4]  + a[2,0]*a[3,1]*a[1,2]*a[4,3]*a[0,4]
         + a[2,0]*a[3,1]*a[4,2]*a[0,3]*a[1,4]  + a[2,0]*a[3,1]*a[4,2]*a[1,3]*a[0,4]
         + a[2,0]*a[4,1]*a[0,2]*a[1,3]*a[3,4]  + a[2,0]*a[4,1]*a[0,2]*a[3,3]*a[1,4]
         + a[2,0]*a[4,1]*a[1,2]*a[0,3]*a[3,4]  + a[2,0]*a[4,1]*a[1,2]*a[3,3]*a[0,4]
         + a[2,0]*a[4,1]*a[3,2]*a[0,3]*a[1,4]  + a[2,0]*a[4,1]*a[3,2]*a[1,3]*a[0,4]
         + a[3,0]*a[0,1]*a[1,2]*a[2,3]*a[4,4]  + a[3,0]*a[0,1]*a[1,2]*a[4,3]*a[2,4]
         + a[3,0]*a[0,1]*a[2,2]*a[1,3]*a[4,4]  + a[3,0]*a[0,1]*a[2,2]*a[4,3]*a[1,4]
         + a[3,0]*a[0,1]*a[4,2]*a[1,3]*a[2,4]  + a[3,0]*a[0,1]*a[4,2]*a[2,3]*a[1,4]
         + a[3,0]*a[1,1]*a[0,2]*a[2,3]*a[4,4]  + a[3,0]*a[1,1]*a[0,2]*a[4,3]*a[2,4]
         + a[3,0]*a[1,1]*a[2,2]*a[0,3]*a[4,4]  + a[3,0]*a[1,1]*a[2,2]*a[4,3]*a[0,4]
         + a[3,0]*a[1,1]*a[4,2]*a[0,3]*a[2,4]  + a[3,0]*a[1,1]*a[4,2]*a[2,3]*a[0,4]
         + a[3,0]*a[2,1]*a[0,2]*a[1,3]*a[4,4]  + a[3,0]*a[2,1]*a[0,2]*a[4,3]*a[1,4]
         + a[3,0]*a[2,1]*a[1,2]*a[0,3]*a[4,4]  + a[3,0]*a[2,1]*a[1,2]*a[4,3]*a[0,4]
         + a[3,0]*a[2,1]*a[4,2]*a[0,3]*a[1,4]  + a[3,0]*a[2,1]*a[4,2]*a[1,3]*a[0,4]
         + a[3,0]*a[4,1]*a[0,2]*a[1,3]*a[2,4]  + a[3,0]*a[4,1]*a[0,2]*a[2,3]*a[1,4]
         + a[3,0]*a[4,1]*a[1,2]*a[0,3]*a[2,4]  + a[3,0]*a[4,1]*a[1,2]*a[2,3]*a[0,4]
         + a[3,0]*a[4,1]*a[2,2]*a[0,3]*a[1,4]  + a[3,0]*a[4,1]*a[2,2]*a[1,3]*a[0,4]
         + a[4,0]*a[0,1]*a[1,2]*a[2,3]*a[3,4]  + a[4,0]*a[0,1]*a[1,2]*a[3,3]*a[2,4]
         + a[4,0]*a[0,1]*a[2,2]*a[1,3]*a[3,4]  + a[4,0]*a[0,1]*a[2,2]*a[3,3]*a[1,4]
         + a[4,0]*a[0,1]*a[3,2]*a[1,3]*a[2,4]  + a[4,0]*a[0,1]*a[3,2]*a[2,3]*a[1,4]
         + a[4,0]*a[1,1]*a[0,2]*a[2,3]*a[3,4]  + a[4,0]*a[1,1]*a[0,2]*a[3,3]*a[2,4]
         + a[4,0]*a[1,1]*a[2,2]*a[0,3]*a[3,4]  + a[4,0]*a[1,1]*a[2,2]*a[3,3]*a[0,4]
         + a[4,0]*a[1,1]*a[3,2]*a[0,3]*a[2,4]  + a[4,0]*a[1,1]*a[3,2]*a[2,3]*a[0,4]
         + a[4,0]*a[2,1]*a[0,2]*a[1,3]*a[3,4]  + a[4,0]*a[2,1]*a[0,2]*a[3,3]*a[1,4]
         + a[4,0]*a[2,1]*a[1,2]*a[0,3]*a[3,4]  + a[4,0]*a[2,1]*a[1,2]*a[3,3]*a[0,4]
         + a[4,0]*a[2,1]*a[3,2]*a[0,3]*a[1,4]  + a[4,0]*a[2,1]*a[3,2]*a[1,3]*a[0,4]
         + a[4,0]*a[3,1]*a[0,2]*a[1,3]*a[2,4]  + a[4,0]*a[3,1]*a[0,2]*a[2,3]*a[1,4]
         + a[4,0]*a[3,1]*a[1,2]*a[0,3]*a[2,4]  + a[4,0]*a[3,1]*a[1,2]*a[2,3]*a[0,4]
         + a[4,0]*a[3,1]*a[2,2]*a[0,3]*a[1,4]  + a[4,0]*a[3,1]*a[2,2]*a[1,3]*a[0,4]
```

A.2 Data File Format: `.counted`

In order to efficiently store coincidence-counting data generated by the DPC-230, we designed a custom binary file format. These files, assigned the extension `.counted`, are structured in records of three 4-byte words. The first word denotes the type of data in the record, and the following two words encode that data, as follows:

First word	Value	Meaning
MAGIC	1337	Identifies the file as being .COUNTED format.
TEMPORARY_FILE	101	Marks the file up as being temporary
STOP_METADATA	102	Marks the end of the metadata header
SCAN_TYPE	103	101: Dip/fringe, 102: Static sample, 103: Scripted scan
SCAN_NSTEPS	201	Number of steps per loop
SCAN_NLOOPS	202	Number of repeated loops in total scan
SCAN_INTEGRATION_TIME	203	Integration time per measurement, ms
SCAN_CLOSE_SHUTTER	204	Whether or not the laser shutter was closed at the end of the scan
SCAN_DONT_MOVE	205	If true, motors were disabled during the scan
SCAN_MOTOR_CONTROLLER	206	Index number of the motor controller used.
SCAN_START_POSITION	207	Motor controller position at start of scan, mm / degrees
SCAN_STOP_POSITION	208	Motor controller position at end of scan, mm / degrees
SCAN_LABEL_NBYTES	250	Length in bytes of a text label, which follows this record
Measurement data		
MOTOR_CONTROLLER_UPDATE	301	Records motor controller index and position.
SCAN_LOOP	302	Loop index
SCAN_STEP	303	Step index
INTEGRATION_STEP	304	Integration step number
STOP_INTEGRATING	305	Written when integration has finished
START_COUNT_RATES	401	Start a list of measured countrates
COUNT_RATE	402	Detection pattern as a binary string, and number of events
STOP_COUNT_RATES	403	End the list of countrates
START_PAUSE	404	Experimentalist paused the measurement
STOP_PAUSE	405	Experimentalist resumed the measurement

Now that the counting system is a little more mature, this format should really be retired in favour of a less opaque standard.

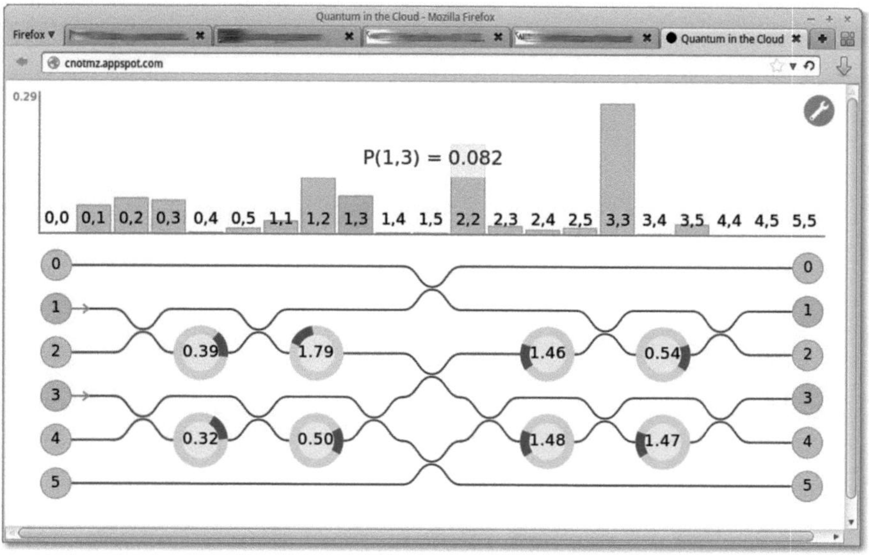

Fig. A.2 Accessible multiphoton simulation of the CNOT-MZ, running in a web browser

A.3 CNOT-MZ API

Much is made in the popular press of the potential impact and power of quantum computing, however the subject is still treated with a certain amount of trepidation, owing to the percieved difficutly of the field, creating barriers to entry for engineers and scientists from other disciplines. In an effort to make quantum computing somewhat more tangible, we built an open-access interface to the CNOT-MZ, accessible through a web browser. Users can run simulations of multiphoton experiments (Fig. A.2), either through a graphical user interface (GUI), or using an hypertext transfer protocol (HTTP) JSON application protocol interface (API). Once granted permission, they can then acquire data from the lab in real-time.

For further detail, see

`https://cnotmz.appspot.com`

Appendix B
Metadata

In writing my thesis, the Ph.D. theses of my colleagues and forbears (notably Jonathan Matthews, Alberto Politi, Alberto Peruzzo, Damien Bonneau, Dylan Saunders and Nathan Langford) have been an indispensable source of detailed, relevant information, clear explanation, and a model for the style and structure of a thesis.

In the hope that it might be useful to other Ph.D. students going through the same process, I include the dataset shown in Fig. B.1. This is a log of approximate word count of my thesis, recorded every time i committed a revision to my `git` repository. Three aspects of this figure are interesting: first, the striking linearity of the curve, which i absolutely expected to be a polynomial with positive second derivative. Second, one can easily identify regions of "burnout" directly after large streaks of progress: I would suggest that this stop-start mode of operation be avoided as far as possible. The last observation, I will leave as an exercise for the reader.

B.1 Dr Peter Shadbolt

Curriculum Vitae

- Web: peteshadbolt.co.uk
- Email: hello@peteshadbolt.co.uk
- Code: github.com/peteshadbolt
- Address: Controlled Quantum Dynamics, Level 12 EEE, Imperial College London

B.2 Academic

- **2015–:** PDRA at Imperial College
  - Theory of measurement-based linear-optical quantum computing
  - Solid-state sources of entangled photons (NV diamond, quantum dot)
  - Quantum algorithms for quantum chemistry

P. Shadbolt, *Complexity and Control in Quantum Photonics*,
Springer Theses, DOI 10.1007/978-3-319-21518-1

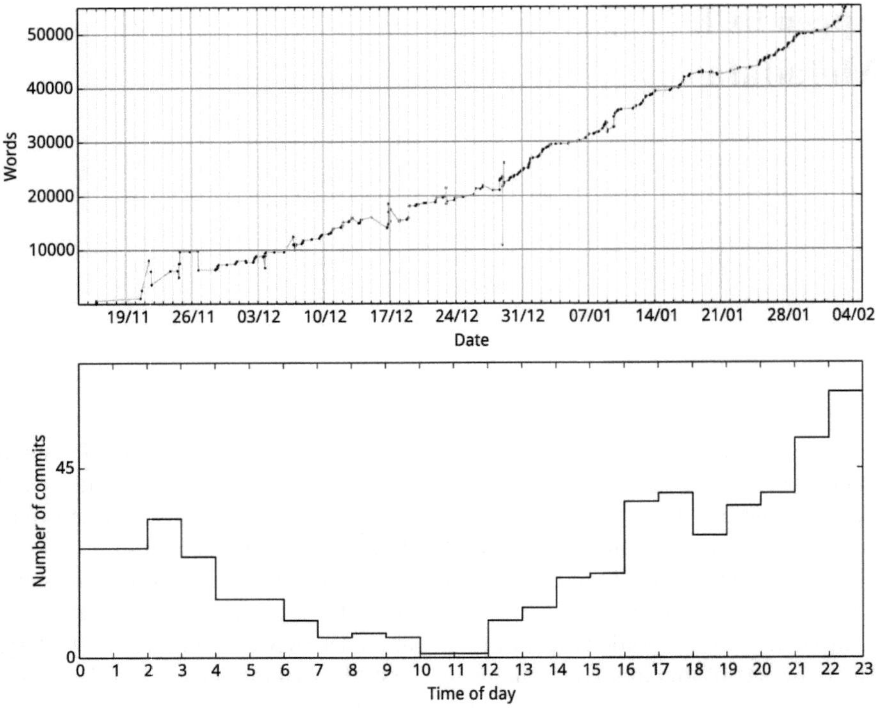

Fig. B.1 Writing a Ph.D. thesis

- **2009–2014:** Ph.D. experimental quantum photonics, University of Bristol
  - Integrated quantum photonics: Generation, manipulation and measurement of entanglement and mixture in silica-on-silicon waveguide circuits.
  - Foundations: reference-frame independent Bell inequalities, a quantum delayed choice experiment.
  - Quantum simulation of Helium Hydride on a photonic chip.
  - Quantum walks and BOSONSAMPLING, generation and measurement of 40,000-dimensional multiphoton states.
  - Construction of a time-correlated single-photon counting system using 16 avalanche-diode photon detectors.
- **2005–2009** MPhys Physics (1st Class Hons), University of Leeds
  - A novel field ionization detector for the micromaser *2008–2009, University of Leeds*
  - Studies of micromechanically exfoliated graphene *2008, University of Leeds*
  - Polymer gel electrolytes for lithium ion batteries *2007, University of Leeds*
- **1999–2005** Royal Grammar School, Buckinghamshire, England
  - 4 A Levels, 11 GCSEs and Advanced Mathematics

B.3 Publications

- P. Shadbolt, M.R. Verde, A. Peruzzo, A. Politi, A. Laing, M. Lobino, J.C.F. Matthews, J.L. O'Brien, Generating, manipulating and measuring entanglement and mixture with a reconfigurable photonic circuit. Nat. Photonics **6**, 45–49 (2012)
- A. Peruzzo, P. Shadbolt, N. Brunner, S. Popescu, J.L. OăĂŹBrien, A quantum delayed-choice experiment. Science **338**, 634–637 (2012)
- P. Shadbolt, J.C.F. Mathews, A. Laing, J.L. O'Brien, Testing foundations of quantum mechanics with photons. Nat. Phys. **10**, 278–286 (2014)
- P.J. Shadbolt, T. Vertesi, Y.C. Liang, C. Branciard, N. Brunner, J.L. O'Brien, Guaranteed violation of a Bell inequality without aligned reference frames or calibrated devices. Sci. Rep. **2**, 470 (2012)
- J. Carolan, J.D.A. Meinecke, P. Shadbolt, N.J. Russell, N. Ismail, K. Worhoff, T. Rudolph, M.G. Thompson, J.L. O'Brien, J.C.F. Matthews, A. Laing, On the experimental verification of quantum complexity in linear optics. Nat. Photonics **8**, 621–626 (2014)
- A. Peruzzo, J. McClean, P. Shadbolt, M.-H. Yung, X.-Q. Zhou, P.J. Love, A. Aspuru-Guzik, J.L. O'Brien, A variational eigenvalue solver on a quantum processor. Nat. Commun. **5**, 4213 (2014)
- H.W. Li, J. Wabnig, D. Bitauld, P. Shadbolt, A. Politi, A. Laing, J.L. O'Brien, A.O. Niskanen, Calibration and high fidelity measurement of a quantum photonic chip. New J. Phys. **15** (2013)
- H.W. Li, S. Przeslak, A.O. Niskanen, J.C.F. Matthews, A. Politi, P. Shadbolt, A. Laing, M. Lobino, M.G. Thompson, J.L. O'Brien, Reconfigurable controlled two-qubit operation on a quantum photonic chip. New J. Phys. **13** (2011)
- J.C.F. Matthews, X.-Qi Zhou, H. Cable, P. Shadbolt, D.J. Saunders, G.A. Durkin, G.J. Pryde, J.L. O'Brien, Practical quantum metrology. arXiv:1307.4673
- M. Gimeno-Segovia, P. Shadbolt, D.E. Browne, T. Rudolph. From three-photon GHZ states to universal ballistic quantum computation (2014). arXiv:1410.3720

B.4 Conference Talks

- Photon10, Southampton (2010) *Poster prize*,
- SPIE Optics and Optoelectronics, Prague (2011) *Invited*,
- Quantum Information Processing and Communication, Zurich (2011) *Invited*,
- Southwest Quantum Information and Technology, New Mexico (2012) *Invited*,
- CLEO 2012, San Jose (2012),
- PRACQSYS 2012, Tokyo (2012) *Invited*,
- QuAMP 2012, Oxford (2012) *Poster*,
- Bristol Cycle Festival, Bristol (2012),
- IOP Quantum technologies: taking concepts through to implementations, London (2012) *Poster*,

- Quantum Optics in the Solid State, Sheffield (2012) *Invited*,
- 13th International conference on squeezed states and uncertainty relations, Nuremberg (2013) *Invited*,
- SPIE Optics and Photonics, San Diego (2013) *Invited* $\times 2$,
- GeneExpression Systems Quantum Science Symposium, Boston (2013) *Invited*,
- Quantum Simulations, Benasque (2013) *Invited*,
- Quantum Simulations and Quantum Walks, Pisa (2013) *Invited*

B.5 Awards

- Photon10 Poster prize
- 2014 EPSRC Rising Star
- 2014 BSA Media Fellow
- Nominated 2015 Springer Theses

B.6 Outreach

- Royal Society summer exhibition *Summer 2011, London*
- Public lecture, bicycle physics *Bristol Cycle Festival 2012*
- Public lecture, nonlocality *@Bristol Science Museum 2013*
- Public lecture, quantum physics *Cafe Kino 2013*
- Web interface to a two-qubit photonic chip: http://www.cnotmz.appspot.com, >40,000 hits

B.7 Personal

Built a machine so that goldfish can play the drums and make drawings. Built a diffuse-illumination multi-touch surface for the Royal Society. Public lectures on bike physics, quantum physics, nonlocality. Black belt WTF Taekwondo. I make electronic music and films, I brew beer. I can hold a conversation in French.

MIX
Papier aus verantwortungsvollen Quellen
Paper from responsible sources
FSC® C105338
FSC
www.fsc.org

If you have any concerns about our products,
you can contact us on
ProductSafety@springernature.com

In case Publisher is established outside the EU,
the EU authorized representative is:
Springer Nature Customer Service Center GmbH
Europaplatz 3, 69115 Heidelberg, Germany

Printed by Libri Plureos GmbH
in Hamburg, Germany